THE CURVE OF BINDING ENERGY

BY JOHN MCPHEE

The Ransom of Russian Art
Assembling California
Looking for a Ship
The Control of Nature
Rising from the Plains
Table of Contents
La Place de la Concorde Suisse
In Suspect Terrain
Basin and Range
Giving Good Weight
Coming into the Country
The Survival of the Bark Canoe
Pieces of the Frame
The Curve of Binding Energy
The Deltoid Pumpkin Seed
Encounters with the Archdruid
The Crofter and the Laird
Levels of the Game
A Roomful of Hovings
The Pine Barrens
Oranges
The Headmaster
A Sense of Where You Are

The John McPhee Reader
The Second John McPhee Reader

John McPhee

THE CURVE OF BINDING ENERGY

FARRAR, STRAUS AND GIROUX

NEW YORK

Farrar, Straus and Giroux
18 West 18th Street, New York 10011

Printed in the United States of America
Published in 1974 by Farrar, Straus and Giroux
First paperback edition, 1980

The text of this book originally appeared
in The New Yorker, *and was developed with the*
editorial counsel of William Shawn and Robert Bingham.

The Library of Congress has cataloged the hardcover edition as follows:
McPhee, John A.
The curve of binding energy.
1. Atomic bomb. 2. Atomic energy industries—
Security measures. 3. Taylor, Theodore B. 1925–
I. Title.
UF767.M215 1974 621.48'3 74-1226

Paperback ISBN-13: 978-0-374-51598-0
Paperback ISBN-10: 0-374-51598-0

www.fsgbooks.com

40 39 38 37 36 35 34 33

To William Shawn

THE CURVE OF BINDING ENERGY

To many people who have participated professionally in the advancement of the nuclear age, it seems not just possible but more and more apparent that nuclear explosions will again take place in cities. It seems to them likely, almost beyond quibbling, that more nations now have nuclear bombs than the six that have tested them, for it is hardly necessary to test a bomb in order to make one. There is also no particular reason the maker need be a nation. Smaller units could do it—groups of people with a common purpose or a common enemy. Just how few people could achieve the fabrication of an atomic bomb on their own is a question on which opinion divides, but there are physicists with experience in the weapons field who believe that the job could be done by one person, working alone, with nuclear material stolen from private industry.

What will happen when the explosions come—when a part of New York or Cairo or Adelaide has been hollowed out by a device in the kiloton range? Since even a so-called fizzle yield could kill a number of thousands of people, how many nuclear detonations can the world tolerate?

Answers—again from professional people—vary, but many will say that while there is necessarily a limit to the amount of nuclear destruction society can tolerate, the limit is certainly not zero. Remarks by, for example, contemporary chemists, physicists, and engineers go like this (the segments of dialogue are assembled but not invented):

"I think we have to live with the expectation that once every four or five years a nuclear explosion will take place and kill a lot of people."

"What we are taking with the nuclear industry is a calculated risk."

"It is simply a new fact of existence that this risk will exist. The problem can't be solved. But it can be alleviated."

"Bomb damage is vastly exaggerated."

"What fraction of a society has to be knocked out to make it collapse? We have some benchmarks. None collapsed in the Second World War."

"The largest bomb that has ever been exploded anywhere was sixty megatons, and that is one-thousandth the force of an earthquake, one-thousandth the force of a hurricane. We have lived with earthquakes and hurricanes for a long time."

"It is often assumed true that a full-blown nuclear war would be the end of life on earth. That is far from the truth. To end life on earth would take at least a thousand times the total yield of all the nuclear explosives existing in the world, and probably a lot more."

"After a bomb goes off, and the fire ends, quiet descends again, and life continues."

"We continue in the direction we're going, and take every precaution, or we go backward and outlaw the atom. I think the latter is a frivolous point of view. Man has never taken such a backward step. In the fourteenth century, people must have been against gunpowder, and people today might well say they were right. But you don't move backward."

"At the start of the First World War, the high-explosive shell was described as 'the ultimate weapon.' It was said that the war could not last more than two weeks. Then they discovered dirt. They found they could get away from the high-explosive shell in trenches. When hijackers start holding up whole nations and exploding nuclear bombs, we must again discover dirt. We can live with these bombs. The power of dirt will be reexploited."

"There is an intensity that society can tolerate. This

means that x number could die with y frequency in nuclear blasts and society would absorb it. This is really true. Ten x and ten y might go beyond the intensity limit."

"I can imagine a rash of these things happening. I can imagine—in the worst situation—hundreds of explosions a year."

"I see no way of anything happening where the rubric of society would collapse, where the majority of the human race would just curl up its toes and not care what happens after that. The collective human spirit is more powerful than all the bombs we have. Even if quite a few nuclear explosions go off and they become part of our existence, civilization won't collapse. We will adapt. We will go on. But the whole thing is so unpleasant. It is worth moving mountains, if we have to, to avoid it."

"A homemade nuclear bomb would be a six-by-six-foot monster. It would take cranes to lift it. You're not going to get a sophisticated little thing that fits into a desk drawer."

"No. But you could get something that would fit under the hood of a Volkswagen."

"If it is possible to build such a device, the situation will come up. We just should be prepared for it, and not sit around wringing our hands. You can't solve this problem emotionally. No. 1: This is a hazard. No. 2: The strictest *practicable* measures have to be taken to prevent it."

"We have to ask ourselves, 'What are we spending our money on, and what are we getting out of it?' I don't believe we can protect ourselves against every bogeyman in the closet. I think we have to take the calculated risk."

Some years ago, Theodore B. Taylor, who is a theoretical physicist, began to worry full time about this subject. He developed a sense of urgency that is shared by only a small proportion of other professionals in the nuclear world, where the general attitude seems to be that there is little to worry about, for almost no one could successfully make a nuclear bomb without retracing the Manhattan Project. Taylor completely disagrees. In the course of a series of travels I made with him to nuclear installations around the United States, he showed me how comparatively easy it would be to steal nuclear material and, step by step, make it into a bomb. Without revealing anything that is not readily available in print, he earnestly wishes to demonstrate to the public that the problem is immediate. His sense of urgency is enhanced by the knowledge that the nuclear-power industry has entered an era of considerable growth, and for every kilogram of weapons-grade nuclear material that exists now hundreds will exist in the not distant future. To give substance to his allegations, he feels he must go into ample detail—not enough to offer an exact blueprint to anyone, to cross any existing line of secrecy, or to assist criminals who have the requisite training by telling them anything they could not find out on their own, but enough to make clear beyond question what could happen.

The source and the reach of his worry result from his own experience. He knows how to do what he fears will be done. Peers and superiors considered him stellar at it once, and used that word to describe him. When he was in his twenties and early thirties, he worked in the Theoretical Division at Los Alamos Scientific Laboratory, where he was a conceptual designer of nuclear bombs. He designed Davy Crockett, which in its time was the lightest and smallest fission bomb ever made. It weighed less than fifty pounds. He designed Hamlet, which, of all things, was the most efficient fission bomb ever made in the kiloton range. And he designed the Super Oralloy Bomb, the largest-yield fission bomb that has ever been exploded anywhere.

WHEN Ted Taylor was growing up, in Mexico City in the nineteen-thirties, he had three particular interests, and they were music, chemistry, and billiards. His father had been a widower with three sons who married a widow with a son of her own, so Ted had four older half brothers—so much older, though, that he was essentially raised an only child, in a home that was as quiet as it was religious. His maternal grandparents were Congregational missionaries in Guadalajara. His father, born on a farm in Kansas, was general secretary of the Y.M.C.A. in Mexico. His mother was the first American woman who ever earned a Ph.D. at the National University of Mexico. Her field was Mexican literature. The spirit of revolution, which had peaked in Mexico long before Ted was born, was still very much in the air, and his earliest impression of politicians was that they were people who carried silver-plated pearl-handled Colt .45s, wore cartridge belts the size of cummerbunds, and went around in Cadillacs firing random shots into crowds of people whose numbers were weighted toward the opposition. Elections, he decided, were a time to stay home. Moreover, politicians were not the only menace in the streets. One time, Ted went out—he was eight—and met a man who told him that he could have a new bicycle if he would go back inside and get something pretty. He went in and got his mother's most precious ring and gave it to the man. Only too late did he realize what had happened, and he burst into tears. He went to the American School, where he started fourth grade one year and finished sixth grade at the end of the same year, thus finding himself about three years younger than most of his friends as he emerged into his teens. In the mornings, before school, he would sit for an hour and listen to music, occupying himself with nothing else while he did so. Years later, he would notice a difference among physicists with regard to music. Working in a scientific enclave at Cornell, where room after room had been equipped with speakers that were connected to a common source of classical music, he found that the theoretical physicists all embraced the music, while the experimental physicists uniformly shut it off. (He also would find that theoretical physicists tended to be loose-knit liberal Democrats, while experimental physicists—conservatives, Republicans—showed a closer weave.) In the afternoons after school, for a number of years, Ted played billiards almost every day, averaging about ten hours of billiards a week. He was, among his friends, exceptionally skillful. He knew nothing of particle physics—of capture cross-sections and neutron scattering, of infinite reflectors and fast-neutron-induced fission chain reactions—but in a sense he was beginning to learn it, because he understood empirically the behavior of the interacting balls on the table, and the nature of their elastic collisions, all within the confining framework of the reflector cushions. "It was a game of skill, dealing with predictable situations—an exact game. The reason it appealed to me was probably the same reason physics appeals to me. I like to be able to predict what will happen and have it come out that way. If you play billiards a lot, you find you can have a great deal of control over what happens. You can get all kinds of things to happen. I have thought of billiard balls as the examples in physics as long as I can remember—as examples of types of collisions from Newton's mechanics to atomic particles. The balls made a satisfying click if they were new and expensive. Downtown, they were new and expensive. It was a treat to go downtown. You could try a twelve-cushion shot there."

He developed a quiet and somewhat shy personality, and considerable self-sufficiency, but he overcame his shyness to dance through long weekends and drink his share of Cuba Libres. Sometimes, he and his friends went off to Acapulco, as many as fifteen teenagers on the loose, and they took one hotel room, for the toilet and the shower, and slept on cots lined up in a long row on the beach. His family lived part of the time in Cuernavaca, which had almost no electricity then (a generator ran the Cuernavaca movie house), and Ted developed there a lifelong preference for candlelight. If the supply-and-demand ratio for electric power were based on him, there would be no power stations, nuclear or fossil. He remembers—almost more than any other image from Mexico—the bread bin, a small wooden box full of bread, in the middle of the table in Cuernavaca, surrounded by burning candles. His thoughts would wander then, as they do now, for remarkably long periods of time, and when he went off into other worlds in Cuernavaca his eyes must have glazed for hours, reflecting the candle flame.

At home in Mexico City—a street-corner house, Atlixco 13—there were certain books that contained pages that could unfailingly cause in him a sensation of terror. They were atlases and geographies, mainly, and he knew just where they were—which book, which shelf.

He would muse, and his eyes would wander to one of them, and he would go and get it. He would open to a picture of the full moon or of a planet—any disclike thing seen in full view—and his flesh would contract with fear. He could never look through a telescope without steeling himself against the thought of seeing a big white disc. He began to have recurrent dreams that would apparently last his lifetime, for he still has them, of worlds, planets, discs filling half his field of vision, filling all his nerves with terror. And yet he could not imagine anything more exciting than having travelled to and being about to land on Mars. He wanted to go there desperately. Years later, he would make intensive preparations to go to Mars in a ship of his design, driven by two thousand exploding nuclear bombs.

When he was ten, he was given a chemistry set for Christmas, and he steadily built it up, year after year, until Atlixco 13 had a laboratory that might have served a small and exclusive university. Things were available from local druggists that would not have been available to him in the United States. Corrosive chemicals. Explosive chemicals. Nitric acid. Sulphuric acid. He enjoyed putting potassium chlorate and sulphur under Mexico City streetcars. There was a flash, and a terrific bang. He made guncotton by the bale. He soaked cotton in nitric and sulphuric acid, thus producing nitrocellulose, then washed it in water, squeezed it, and hung it up to dry. The result looked just like cotton but would explode—*poof*—and leave almost no ash. It was pretty at night. He once wadded it into a .22 cartridge and hit the cartridge with a hammer. The cartridge went into his finger. He hunted through the 1913 New International Encyclopaedia, which contained lots of chemistry, and he found many things to make. He made urethan (ethyl carbamate), a sleep-inducing drug, starting from a point very close to scratch. He first needed urea, and the nearest source was his own bladder, so he drained it out and went to work. He boiled a pint of urine until he had a half cup, then precipitated out the urea. He added nitric acid, and got urea nitrate. He added formaldehyde, and got crystals of urethan. He tried it on his white rats, and put them to sleep for up to twelve hours at a time, but he brought the dose up slowly, and he killed no rats. He worked in his chem lab three hours a day in term, and all through the annual long vacations, which came in winter and lasted two and a half months. He liked the beauty of some precipitates, and the most beautiful by far, he thought, was lead iodide. It looked like gold dust being sprinkled into water when, with light behind a beaker, he dropped lead-acetate solution from an eyedropper into sodium iodide. Particulate flakes of gold drifted down, shimmering, sparkling with gold light. He made a yellow-and-red powder that was a combination of picric acid and red lead. It was a relatively stable material, but it would detonate, given sufficient heat. He would set a little pile of it on a piece of one-sixteenth-inch steel plate and heat the plate from below. Flash. Bang.

One-quarter teaspoon of the mixture, unconfined, would blow a hole right through the steel. In repeated experiments, he figured out exactly how little powder was needed to penetrate the plate. He added ammonia to a concentrated solution of iodine crystals in alcohol. The resulting precipitate, filtered out, was a wet, blackish blob of nitrogen iodide. He dried it. Dry nitrogen iodide is stable with regard to heat but unstable with regard to motion. It can literally be exploded by tickling it with a feather. Ceilings were high in Mexico, and there were long feather dusters at Atlixco 13. Holding one like an épée, Ted would reach gingerly toward a mound of nitrogen iodide. Flash. Bang. A purplish-brown cloud. A miniature mushroom. His mother was incredibly tolerant of his chemical experimentation. He was graduated from high school when he was fifteen.

THE material that destroyed Hiroshima was uranium-235. Some sixty kilograms of it were in the bomb. The uranium was in metallic form. Sixty kilograms, a hundred and thirty-two pounds, of uranium would be about the size of a football, for the metal is compact—almost twice as dense as lead. As a cube, sixty kilograms would be slightly less than six inches on a side. U-235 is radioactive, but not intensely so. You could hold some in your lap for a month and not suffer any effects. Like any heavy metal, it is poisonous if you eat enough of it. Its critical mass—the point at which it will start a chain reaction that will not stop until a great deal of energy has been released—varies widely, depending on what surrounds it. If the uranium is wrapped in steel, for example, its critical mass is much lower than it would be if the uranium were standing free. A nuclear explosion is a chain reaction that goes so fast that pressures build up in the material and blow it apart. Depending on the capabilities of the designer, a given mass of U-235—say, twenty kilograms, an amount slightly smaller than a grapefruit—can yield an explosion equivalent to anything from a few tons of TNT up to hundreds of thousands of tons of TNT (hundreds of kilotons). The bomb of Hiroshima, which was not efficiently designed, fissioned only one per cent of its uranium and yielded only thirteen kilotons. There are various ways to make nuclear bombs, some of which require less material than others. It is theoretically possible to make a very destructive bomb with nuclear material the size of a pea, but that is beyond the practical capability of even the man of extraordinary skill in the art. Musing once over a little sliver of metallic U-235 about the size of a stick of chewing gum, Ted Taylor remarked, "If ten per cent of this were fissioned, it would be enough to knock down the World Trade Center." The United States

Atomic Energy Commission has set five kilograms of U-235 as the amount at and above which the material is "significant." A bomb might be made with less. A crude bomb would require more. Five kilograms is an arbitrarily chosen figure—an amount which, if stolen, would be cause for concern. The Atomic Energy Commission, now much occupied with the growth and development of the peaceful nuclear-power industry, wants the atom to make a good impression on the general public. In the frankly bellicose days of the somewhat forgotten past, the term used was not "significant" but "strategic." Unofficially—around the halls and over the water coolers—five kilos is known as "the trigger quantity."

Uranium as found in nature, and mined, and milled, and extracted as metal, is worth about twenty-five dollars a kilogram. Uranium-235 is worth as much as twenty thousand dollars a kilogram. The reason for this great difference is that for each atom of U-235 that exists in natural uranium there are a hundred and forty atoms of U-238. It is U-235 that makes the fissions, makes the bombs, makes the heat in the power reactors; and the U-235 is extraordinarily difficult to separate from the rest of the uranium. A uranium atom—any uranium atom—has ninety-two protons: spheres bunched up in its nucleus. In there, too, like so much additional caviar, are many neutrons—a hundred and forty-six neutrons in an atom of U-238 (92 + 146 = 238), and a hundred and forty-three neutrons in an atom of U-235. Separating these two sisters, these two isotopes, was one of the hardest things human beings have ever tried to learn how to do, because, for one thing, U-235 and U-238 behave chemically in an identical way. So the isotopes had to be separated physically. It was necessary for the people who were trying to do this to get down on their knees, in effect, and sort into piles tiny spheres whose diameters were expressible in hundred-millionths of centimetres and whose only difference was that one kind weighed ever so slightly more than the other. This became, and has essentially remained, the most secret aspect of the development of nuclear material. Various methods were tried. The most cumbersome and, at least until recently, the most effective method was gaseous diffusion. Natural uranium was combined with fluorine and turned into a gas: uranium hexafluoride, UF_6. The gas was sent drifting through incredibly thin membranes. No one is saying exactly what the membranes consisted of, but they were successfully created and they are still in use. The gaseous-diffusion process was necessarily marginal in its efficiency. Both kinds of molecules went streaming through the membranes, but because the U-235 atoms were a little over one per cent lighter, and therefore were moving faster, a little extra U-235 went through in any given pass, and the uranium on the other side of the membrane was, as the technologists put it, enriched. The enrichment was so very slight, though, that the process had to be repeated again and again. The gas had to flow through several thousand membranes, which, cumulatively, became known as "the cascades." Thousands of miles of tubes, pipes, and other conduits were needed to create a network of flow wherein the gas could now go through a membrane, now return to try again, now go on to a new membrane, gradually advancing, in a process of separation and elimination, until what had begun as seven-tenths of one per cent U-235 was more than ninety per cent U-235—fully enriched, weapons-grade uranium.

Gaseous-diffusion plants cover hundreds of acres. They are so big that people drive automobiles and ride bicycles inside them, down long corridors among the cascades. There are three in the United States, all operated under A.E.C. contracts: one in Oak Ridge, Tennessee; one in Portsmouth, Ohio; and one in Paducah, Kentucky. At least four more enrichment plants must be built in the United States alone before 1985. Nothing about them is cheap. It takes a big power plant—enough to serve a city—just to run one gaseous-diffusion plant. The existing ones get their energy from power plants that burn strip-mined coal. Some people used to wonder aloud when the nuclear industry was going to produce more power than it was using—a question that was regarded by the industry as "a sick joke." Two billion dollars will buy a gaseous-diffusion plant. Britain has one. France has one. Needless to say, the Russians have however many they need. When the Chinese exploded a uranium bomb in 1964, it was assumed that the Chinese were not smart enough to have figured out the technology of isotopic separation. Therefore, the Chinese must have stolen the U-235. Where? No one could guess. Some months later, though, it was disclosed that sixty kilograms of U-235 was unaccounted for at a nuclear-fuel-fabricating plant in Apollo, Pennsylvania. Perhaps the Chinese had stolen the uranium in Pennsylvania. While this speculation was going on, the government revealed that a reconnaissance plane had made a high overflight above China and taken photographs that showed the presence of a gaseous-diffusion plant at Langchow, in Kansu Province.

The complexity of gaseous diffusion has importantly helped to confine the spread of nuclear weapons. Anybody could get hold of uranium, but it was another matter to get hold of a gaseous-diffusion plant. The development of other methods of isotopic separation has weakened that barricade, and there is a possibility now that it has broken down altogether. When prospectors screened ore for gold, they were doing something analogous to the gaseous-diffusion process. When they panned for gold, though, they did something quite different: they put an ore slurry in a pan and took what settled. Uranium isotopes can be separated that way, too—in centrifuges whirling around and flinging the heavier U-238 to the outside. There is so little U-235 to begin with that this also is a long and clumsy process, involving tens to hundreds of thousands of centrifuges;

but, at least theoretically, it takes less power and less space, and since the centrifuges can be spread out geographically and not contained in one plant, a country (if not a group of people) could the more easily enrich uranium in a secret operation. South Africa has announced that it has developed an entirely new way of enriching uranium but will not give any clue to what it is, whether it is liquid or gaseous, centrifugal or centripetal, white or black. Simplest of all, in terms of space and equipment required, is a method under development by, among others, a team of American physicists, who have reported various approaches to separating uranium isotopes with a laser. If that proves possible, several skilled individuals could do it almost anywhere if they could assemble the right equipment. Thus, all the uranium on the near side of the enrichment plant—in the mine, in the mill, in the factory that turns it into UF_6—may soon be vulnerable to misuse. Meanwhile, in an attempt to serve the burgeoning growth of the nuclear-power industry, and to solve some of the economic problems attendant upon it, the Atomic Energy Commission has announced that it is giving up its enrichment monopoly and that it is going to license private corporations to build gaseous-diffusion plants of their own.

Almost all power-plant reactors now making electricity for home use do not use fully enriched uranium. In their fuel elements (also called fuel assemblies), they use uranium that has been enriched only until it is about three per cent U-235. This includes the entire present generation of so-called light-water reactors, all the "nuclear plants" that belong to Consolidated Edison Company of New York, Connecticut Yankee Atomic Power, Jersey Central Power & Light, Pacific Gas & Electric, and so forth. A nuclear bomb could not be fashioned from the slightly enriched uranium that goes into their cores, nor could a nuclear explosion occur as a result of some sort of error or accident at such a plant. Where, then, is the more than half a million kilograms of weapons-grade uranium that has been produced in the United States since 1945? Roughly two per cent has been exploded. Something less than that has been consumed in various small reactors that use fully enriched uranium. Most of the rest is strewn around the world in the cellars and silos of the military-weapons program, in the form of bombs. Nuclear submarines burn fully enriched uranium. Several kinds of small test reactors—sold by American companies to nations and universities all over the world—use fully enriched uranium. There is a new kind of power reactor, known as the H.T.G.R., that uses a great deal of fully enriched uranium and is so promising that it may one day predominate over the type now in use.

The only American diffusion plant now producing fully enriched uranium is at Portsmouth, Ohio. The material, UF_6 in solid form, is shipped from Portsmouth in ten-litre steel bottles, generally by airplane or truck, to conversion plants that turn it into uranium oxide or uranium metal—whichever the customer wants. Each ten-litre bottle contains seven kilograms of U-235, and there may be, typically, twenty bottles in a shipment. Conversion facilities are in Hematite, Missouri; Apollo, Pennsylvania; Erwin, Tennessee. Then the oxide or the metal is shipped on, again by air or truck, to fuel-fabrication plants, which are in Crescent, Oklahoma; New Haven, Connecticut; San Diego, California; Lynchburg, Virginia. The oxide is a fine brown powder that looks like instant coffee. The metal comes in small chunks known as "broken buttons." (William Higinbotham, a physicist at the Atomic Energy Commission's Brookhaven National Laboratory, says that to fashion a nuclear explosive from broken buttons "all you'd have to do is hammer it into the right form and you're ready to go.") As oxide or metal, the material travels in small cans that are placed in a cylinder—a five-inch pipe—that is braced with welded struts in the center of an ordinary fifty-five-gallon steel drum. It is for criticality reasons that the uranium is held in the center with the airspace of the drum around it, for if too much U-235, in any form, were to come too close together it would go critical, start to fission, and irradiate the surrounding countryside. The fifty-five-gallon drums with interior weldings are called birdcages, because in a vague way they resemble them. Loaded, they weigh a hundred pounds and can be handled by one person, easily by two. The ten-litre bottles of UF_6 travel in birdcages as well. Because of the criticality danger, such drums are clearly labelled "FISSILE MATERIAL" or, synonymously, "FISSIONABLE MATERIAL." Anybody familiar with the labelling practices of the industry can tell that the contents are of weapons grade.

One place where nuclear-submarine fuel is made is on the corner of Gibbs and Shelton Streets in one of the less expensive neighborhoods of New Haven. The zoning is mixed there. Private homes and apartments are across the street from the plant—United Nuclear. The housing is sort of decayed. Turnover is frequent, many signs in the windows—"FOR RENT, 82 Shelton Street, 624–1200"; "FOR SALE, 33 Gibbs Street, Gatison Lenward Associates, Realtors, 562–2187." Across the street, at any given time, is about a thousand kilograms of U-235, in metallic form, as pure uranium or as uranium-aluminum fuel plates—strips of metal, easily portable, each like the plate a doctor might screw to a door to announce his presence within. In any fabricating operation, there is considerable scrap, and no one is going to throw away something worth many thousands of dollars. So there are half a dozen scrap-recovery plants in the country, and United Nuclear's, for example, is in Wood River Junction, Rhode Island, where birdcages containing about a thousand kilograms of U-235 go in and out each year. They sit outdoors waiting to be reprocessed. The New Haven plant consists of several buildings, one or two as shabby as the tenements opposite. The plant is surrounded by a chain-link-and-barbed-wire fence except where certain walls actually abut the public sidewalk. "New Haven is not well alarmed. You could get through that wall easily," Higinbotham once observed. The uranium that is not actually being processed is stored in a vault. About six hundred and seventy workers come and go in the plant.

In Erwin, Tennessee, fully enriched uranium scrap from the military-weapons program is recovered.

At its Cimarron facility in Crescent, Oklahoma, Kerr-McGee does scrap recovery and also fabricates experimental reactor fuel, handling about five hundred kilograms of U-235 a year.

In Hematite, Missouri, General Atomic has about seven hundred kilograms of U-235 on the premises at any given time, first in the form of UF_6, in bottles from Portsmouth, Ohio, and then, most notably, in oxide form prepared for shipment. Like metallic U-235, the U-235 oxide could be used in a bomb. Some people argue that this is not so—engineers, executives, people in the business—but if they were to carry the argument far enough they would have to argue with Ted Taylor. A great deal of fully enriched uranium oxide has travelled from Hematite to Kansas City in an ordinary common carrier (a truck), then on to Los Angeles as air cargo, then a hundred and twenty miles down the freeways in another ordinary truck to General Atomic, in San Diego. Something over a thousand kilograms that has come to San Diego this way has been turned into particles of uranium dicarbide, mixed with thorium dicarbide. In very small amounts, the mixture was encased in coatings of pyrocarbon and silicon carbide, making beads the size of pinheads. The beads were used to fill holes that had been drilled in blocks of graphite thirty inches high. Roughly ninety blocks at a time (sixteen truckloads) travelled through California, Nevada, Utah, Wyoming, and finally to the Fort St. Vrain power station near Platteville, Colorado, where they were piled one atop another, held together by gravity and small dowels, in the innermost chamber of the H.T.G.R.—the High-Temperature Gas-Cooled Reactor. This new variant has been called "the reactor of the nineteen-eighties." It differs vastly from the present generation. The fuel elements of present light-water reactors, for example, typically consist of long, thin rods of zirconium alloy packed with uranium oxide and sealed at the ends. The graphite blocks of the H.T.G.R. are something new in cost, efficiency, fissions per dollar. The High-Temperature Gas-Cooled Reactor is thrifty with neutrons. It uses less uranium per megawatt. Most reactors operate at six hundred degrees Fahrenheit. The H.T.G.R. functions at fourteen hundred degrees Fahrenheit, a difference that obviously bears a payload, since heat (that makes steam that drives turbine generators that make electricity) is what all power reactors exist to produce. Property of the Public Service Company of Colorado, the H.T.G.R. contains a little over a thousand kilograms of fully enriched weapons-grade uranium. Southern California Edison has ordered two that are twice as big. Philadelphia Electric has ordered two H.T.G.R.s, each three times the size of the one at Fort St. Vrain. Delmarva Power & Light has ordered two like the ones for Southern California. A Japanese industrial complex is considering one for the heat alone.

It would take an impressively sophisticated individual or group to achieve a nuclear explosive starting with the beads in the graphite of the H.T.G.R. The task would be at best laborious and difficult, for the pyrocarbon and silicon-carbide coatings were designed to withstand the temperatures and the pressures in the fissioning core of a reactor named for the high heat within it. So the H.T.G.R. itself is not particularly vulnerable to theft by potential bombmakers. What is relevant is that the H.T.G.R. uses great quantities of weapons-grade uranium, a sixth of its core will be replaced each year, and in order for the U-235 to make its way to the reactor it first has to travel in far more "significant" form from Ohio to the conversion plant in Missouri (or another one somewhere else) and then on to California. The H.T.G.R. is such a good reactor that the volume of this flow to new fabricating plants will before long be in the tens of thousands of kilograms, all over the United States. When that era arrives, a few kilograms will still be the trigger quantity.

General Atomic in San Diego, where the High-Temperature Gas-Cooled Reactor was conceived and developed, is a beautiful complex of buildings—two dozen or so, spaced over many acres—designed by Pereira and Luckman and landscaped with tiered pools, a footbridge, hibiscus, oleander, and jasmine. It stands in open country a few hundred yards from access to Interstate 5. Anyone who wanted to know the general layout of the place, to learn the whereabouts of the vaults in which the uranium is stored, or to learn when shipments might be expected to come and go need not infiltrate the plant or fake the requisite badges or wear a mask and a cloak. It is necessary only to go to a public reading room that the A.E.C. maintains at 1717 H Street in Washington, D.C. A card catalogue there contains General Atomic's docket numbers. A clerk in an adjacent document room waits behind a kind of Dutch door, and a request for any docket number quickly yields a huge stack of papers that contains, among other things, the General Atomic license, in which are diagrams of the plant, capacities of the vaults. There is also voluminous correspondence about future plans (General Atomic is going to build another fuel-fabrication plant, near Youngsville, North Carolina) and about present shipments—more than enough for an analysis of material flow. Similar papers on all nuclear facilities in private hands are available at H Street. The Atomic Energy Act, as amended in 1954, says the public has the right to know about the private use of nuclear materials. H Street is one place where that right can be exercised. A Xerox machine is there for the reader's convenience.

One vault at General Atomic is about thirty by thirty

feet and contains stacks of shelves on which are coffee cans (that is what they are called, anyway, and they are that size) clearly labelled to show the amount of U-235 within—generally about three kilograms per can. The vault's capacity is nine hundred and ninety kilograms, but five hundred is about as much as it ever contains. A vault man in white coveralls is the only person on any shift who can open the combination lock on the door. When the uranium moves out of the vault and around the plant, the coffee cans are set on rolling "move carts"—six cans on a cart. During a visit that Ted Taylor and I made there one day, three move carts, with fifty-four kilograms of U-235 on them, were standing near a big garage-type door that was open to the sunshine outside. We went through the door and found a triple fencing system and a guard in a small guardhouse. Three gates were open in the three fences, and an unmarked pickup truck came in and zipped past the guard, who was resting his chin on his hands and did not look up. "The vault has an intrusion alarm," the plant manager told us. "A big bumblebee will set the son of a bitch off. If someone came in here and started shooting, though, he could get whatever he wanted. One man with the right attitude could do it. But what he was carrying out wouldn't be worth a damn to him. Not that stuff. You can't make a bomb out of that stuff."

IN the fall of 1941, Ted Taylor went to Exeter, for one year of additional secondary schooling, and in the New Hampshire winter he knew frozen ponds and rivers for the first time. He learned to skate. The feel of it energized him in the way that someone else his age might have been excited by a first chance to drive a car. He would skate alone in the afternoons ten miles up the Exeter River, through boggy woods, watching through ice as clear as window glass the rocks and pine needles on the bed of the river. He was taking "Modern Physics" under Elbert P. Little, a teacher of such ability that old Exonians thirty and forty years away from Exeter still remember him with particular and affectionate awe. He gave Ted a D, a flat and final D, and even in the winter term Ted could see that D was his status, and that it was unlikely to rise. He barely noticed, because with his D he was getting a look for the first time—and a vividly clear one—at what he would call "submicroscopic solar systems," and he found that they had for him enormous appeal. One proton with an electron (about eighteen hundred and fifty times lighter) orbiting around it—hydrogen. One proton and one neutron together in a nucleus with an electron orbiting around it—heavy hydrogen (deuterium). Two protons and a number of neutrons with two electrons orbiting around them—helium. Three protons, some neutrons, three whirling electrons—lithium.

One at a time, add a proton and an electron, and each element became another. Four protons, four electrons—beryllium. Five—boron. Six—carbon. Seven—nitrogen . . . Seventy—ytterbium . . . Seventy-eight—platinum. Seventy-nine protons—gold. Eighty protons—mercury (eighty protons massed together with anywhere from ninety-nine to a hundred and twenty-six neutrons into a body around which orbited eighty electrons, whose negative charges exactly balanced the eighty positive charges of the protons). Eighty-one protons—thallium. Eighty-two protons—lead. Bismuth. Polonium. Astatine. Radon. Francium. Radium. Actinium. Ninety protons—thorium. Ninety-one protons—protactinium. Neutrons had no charge and were neither attracted nor repelled by electrical forces and were thus the particles that could most easily be taken out of one atom and shot into the nucleus of another. Ninety-two protons, ninety-two electrons, a gross (more or less) of neutrons —uranium. The list, at the time, stopped there, having included everything that was found in nature. The transuranium elements were just beginning to be discovered and were not known in Exeter. Out on the river, skating, he pondered the root simplicity that all things he had ever seen—wood and water, bread and candle wax—were made of neutrons, protons, and electrons, separated by space. He tried to imagine what it would be like to live on an electron. What would the nucleus look like as a sun? There were a sextillion protons, a sextillion electrons, and a sextillion neutrons in one dead leaf on the bottom of the river. There was an island universe in a drop of water. His imagination outgrew his chemistry lab in Mexico City. He decided that he wanted to be a physicist.

At Exeter, he also learned to throw the discus. He was attracted to the shape and the flight of the thing ("It was the first and last sport in which in any sense I ever excelled"). He was a discus thrower in college as well—at the California Institute of Technology. Cal Tech was a dull and heavy grind for him. He lightened it somewhat by making nitrogen iodide, the stuff he liked to tickle in Mexico; and he would put it wet into the keyholes of the doors of friends who were off on weekends. The material would dry in there and become unstable with respect to motion. A friend would return to Cal Tech and put his key in his lock. Flash. Bang. The explosion was so designed that it would hurt neither the lock nor the man with the key. Charles Cutler, Ted's roommate from those days, has said that what he remembers most about Ted as a college student is how self-contained he was. If they were walking along together and Cutler stopped to tie his shoe, Ted kept right on walking. He never seemed to notice. Cutler developed a similar set of responses, and when Ted stopped to tie one of his own shoelaces Ted stopped alone. They got along fine. Cutler is now Ted's attorney in Washington.

Ted spent his second and third years at Cal Tech in the Navy's V-12 program, accelerating the grind, cramming physics, graduating in June of 1945. He was nineteen. He was sent to midshipman's school at Throgs Neck, in the Bronx, and he was there all summer. He began a letter home on August 8, 1945, but went on shore leave to New Jersey and did not finish the letter until August 13th:

DEAR FOLKS,

. . . Things have been happening so fast the last twenty-four hours that everyone is in pretty much of a daze. I'm on the off-section of the watch now and have some time to take it easy and try to let what's happened sink in.

The headlines about the success of the atom bomb are undoubtedly the biggest news of the century, if not an announcement of the most important single event in the history of the world. My first reaction to the news was one of almost horror, in spite of the fact that I think the end of the war is a matter of weeks. We've been on the threshold of discoveries enabling man to utilize the unlimited energy released by "exploded" atoms for several years, but I never dreamed that the first experiments would be so spectacularly successful—and so destructive. The effective destruction of an entire city by one bomb was unthinkable before the destruction of Hiroshima. Now it is quite possible that Japan may be literally wiped off the map if she doesn't surrender soon.

Some of the revolutionary changes in our industrial systems which will be possible soon are obvious; my fear is that man has discovered something which, knowing so very little about it, may destroy him. This discovery will undoubtedly be common knowledge to all the governments of the world before long. Therefore, it seems to me that this must either be the last war or the nations of the world will completely destroy each other. And it will only be through radical changes in the world's economic, social, and political systems that a complete catastrophe can be prevented. Midshipman's school seems insignificant in the face of what's happened, but we must plug on! We were sworn in yesterday, got the rest of our uniforms and devices, and I must say I'm relieved, although I had practically no doubts about making it. Forty of the three hundred who started were bilged. . . .

I met Bob and Rosemary at Grand Central Saturday afternoon, and we drove back to Plainfield that night. We spent most of the time hashing out the stupendous events of the week.

I called Cousin Mary, but no one answered, so I judge she's gone to the shore for the summer.

. . . I don't suppose the world situation has ever been so critical. With the proper leadership and cooperation of the United Nations, we could very easily be entering the greatest period of progress in our history. The atom bomb and what it represents may easily be the means to end all wars.

As for the effect of the end of the war on me, I can only say that I'll probably have to serve the remaining two years I enlisted for in the Navy. I'm going to pull all the strings I can to get into naval research in atomic physics, now that recent developments have positively obviated the applications of atomic physics to warfare. At any rate, whatever I do the next two years, I'll certainly be all set when I'm discharged. It was certainly a break to have started in early in a field which will undoubtedly be the most important scientific one for many years. . . .

Lots of love,

TED

Listening to the radio in the Throgs Neck (Fort Schuyler) barracks, he had heard the announcement of the destruction of Hiroshima. The bomb was a total surprise to him. The announcer went on to tell about a great flash of light that had been seen in the American Southwest less than a month before, from a bomb that had "vaporized" a tower near Alamogordo. In the weeks that followed, Ted, of course, read every available line of gleaned scientific reportage. He had never before heard the word "fission." When he saw his mother, he said that he felt even more strongly than he had at the time he wrote his letter that he would under no circumstances ever work on a nuclear explosive. In reading accounts of the Alamogordo and Nagasaki bombs, he also encountered for the first time in his life the word "plutonium."

THE material that destroyed Nagasaki was plutonium-239. Plutonium was the first man-made element produced in a quantity large enough to see. It was created in 1940 at the University of California at Berkeley. The idea of it had for many years been indicated by the periodic table of the elements, where a row of blanks paralleling the rare earths suggested the theoretical possibility of elements whose family characteristics—like the characteristics of thorium, protactinium, and uranium—would be similar to those of actinium. It was possible that unknown elements (with ninety-three protons, ninety-four protons, and so on) had long ago existed in our solar system but had vanished because of instability. A stable element is one that lasts a relatively long time compared to the age of the universe—now thought to be around ten billion years. So in order to "discover" the transuranium elements it was necessary to make them, and one way of doing this was neutron capture. When a free neutron enters the nucleus of an atom of uranium-238, for example, the atom becomes uranium-239. Now begins a struggle for stability within the atom, which has been made unstable by the new ratio of protons to neutrons. Spontaneously, a neutron changes into a proton and a new electron is created in the nucleus and goes shooting out with a concomitant display of energy. In this way, one element becomes another. In this instance, it takes about twenty-three minutes for half of a given quantity of uranium-239 (ninety-two protons, a hundred and forty-seven neutrons) to decay into neptunium-239 (ninety-three protons, a hundred and forty-six neutrons), and another twenty-three minutes for half of the remainder to change, and so on. Hence it is probable that any new atom of uranium-239 will have changed into neptunium-239 within an hour. Neptunium-239 is also unstable, and it repeats the process exactly, spontaneously changing a neutron into a proton and creating a new electron. In a day, or two, or three, the atom has become plutonium-239 (ninety-four protons, a hundred and forty-five neutrons), a relatively stable isotope, with a half-life of twenty-four thousand three hundred and sixty years. This sequence of events is happening continuously in all the nuclear power plants now operating in the world, for plutonium is an inherent by-product of the fissioning of uranium in a nuclear reactor.

In great secrecy during the Second World War, big "production reactors" were built at Hanford, Washington, to fission uranium and produce plutonium-239, because, among other things, it had been calculated that two to three times less plutonium-239 than uranium-235 would be required in the making of nuclear explosions. Production was slow at first—grams a day, extracted chemically from spent fuel, in the form of plutonium nitrate in a water solution. As the war itself moved slowly along—from Leyte Gulf to Iwo Jima to Okinawa—the plutonium that would level Nagasaki literally dripped into bottles in Hanford, Washington. One day in 1944, shortly after the first of the reactors had been started up, a balloon appeared overhead in

Hanford. It had been made in Japan, equipped with an incendiary device, and released into the wind. Many hundreds of balloons like this one had travelled all the way across the Pacific, and some had landed in the forests of the American Northwest, where they started fires of considerable magnitude. The fire balloons were so successful, in fact, that papers were asked not to print news of them, because the United States did not want to encourage the Japanese to release more. The balloon that reached Hanford had crossed not only the Pacific but also the Olympic Mountains and the alpine glaciers of the Cascade Range. It now landed on an electric line that fed power to the building containing the reactor that was producing the Nagasaki plutonium, and shut the reactor down.

Once, in the early nineteen-forties, all the plutonium in the world was in a cigar box in a storeroom next to the office of Glenn Seaborg, one of the element's four discoverers. After the advent of privately owned nuclear-power reactors, the United States government bought up—for nine dollars a gram—the plutonium they produced. The policy of buying it from the private power companies was ended, though, in 1970, and since then, for various economic reasons, the companies have been stockpiling their own plutonium. Private companies will soon own more plutonium than exists in all the bombs of NATO. With the predictable growth and expansion of the nuclear industry, power companies will make a cumulative total of ten million kilograms of plutonium within the last quarter of the twentieth century.

The trigger quantity is two kilograms. Enough plutonium to make a bomb could be carried in a paper bag.

The privately owned plants built to perform the chemical separation of plutonium from used nuclear fuel are near West Valley, New York, and Morris, Illinois. The West Valley plant has shut down for improvements and will not reopen for several years. The plant at Morris is a new one.* Fuel elements from reactors come to these places in heavy steel casks on the beds of trucks and railway cars. The casks are lowered into deep, clear, demineralized water, and are opened down there by mechanical arms. The fuel assemblies are removed and are stored on end. A rich purple aurora glows several feet into the water around them—a radiation phenomenon known as the Cherenkov effect. Electrons that are created in the nuclei of decaying atoms are called beta rays, and it is these that emit the purple light. Beta rays are one of three forms of energy that have been grouped under the name radioactivity. The other two, also resulting from nuclear decay, are alpha particles and gamma rays. The deep water completely holds in the intense radioactivity, and people can walk around the edges of the storage pools with impunity. Should anyone fall in, there are ring-buoy life preservers on the walls.

Hydraulic arms move the fuel assemblies, one at a

* Since this book was first published, the Illinois plant encountered technical difficulties of apparently insuperable nature, and it is not in operation. Nonetheless, the description here is generally applicable to any commercial reprocessing plant.

time, into a concrete labyrinth, where they are lifted out of the water and are moved on into a long central room called "the canyon." The walls of the canyon are five feet thick, and contain windows that cost twenty-five thousand dollars apiece. If you hold a match before one of these windows, you see eight flames. The window, five feet thick, is made of eight panels of glass, with mineral oil between the panels. The glass itself contains considerable lead. Each window weighs twelve and a half tons. Inside the canyon, master-slave manipulators perform tasks no human being could do directly without dying almost then and there. Fuel rods are chopped up, like stalks of celery under a kitchen knife, by a mighty guillotine that cleanly severs steel. Now all the small cuttings are dissolved in nitric acid, and the resulting radioactive fluid begins to travel up and down through a series of tanks and tubes, arranged in tall columnar form, where the addition of reactive chemicals—tributyl phosphate, dodecane—effects the separation of plutonium and unconsumed uranium not only from each other but also from a variety of radioactive fission products, such as strontium-90, krypton-85, and cesium-137. At the far end of the canyon, uranium hexafluoride comes out through a hole in the wall. Plutonium-nitrate solution pours from a nearby spigot.

General Electric owns the chemical-reprocessing plant at Morris. It was built on a small plot of ground among fields of corn and soybeans not far from the confluence where the Kankakee and the Des Plaines form the Illinois River. Only sixty miles from Chicago, the plant is nonetheless surprisingly remote. Only about fifteen people are needed to run it, although that is the figure for nighttime ghost-shift operation, and more are there by day. By license requirement, the plant has a telephonic connection with the Illinois State Police and a radio connection with the Grundy County sheriff. The telephone lines are buried. The state police have assured General Electric that a distress call would bring a police car to the scene within fifteen minutes, another car, if needed, within half an hour, and thirty-eight more cars within two hours. At the plant, a guard is always on duty. He is supplied by a company called Advance Industrial Security. On his shirt are an American emblem and a white eagle. On a day when Taylor and I visited the plant, the incumbent on duty was a big man with a ruddy face and a huge belly. He appeared to be the sort of man who could run a hundred yards in four minutes.

The plant was surrounded by a seven-foot chain-link fence topped with barbed wire. Its doors were locked. Its more sensitive areas—for example, the plutonium-load-out cell, the plutonium-storage corridor, the laboratory (where samples are withdrawn from the canyon for assay)—were watched by fourteen closed-circuit-TV cameras, which, in turn, were monitored on two screens in the central control room. The plutonium from the spigot—no longer contaminated with radioactive fission products—goes into slim stainless-steel flasks, about four feet tall, with a five-inch outside diameter. Each flask contains ten litres of plutonium-nitrate solution—roughly two and a half kilograms of plutonium. Richard Fine, a chemist who had worked at Hanford in 1944 and had stayed twenty years before coming to Morris, said to us, "Plutonium should be better protected than money. That fence out there could be gotten over easy as pie. You could back a car up to it. Along the back side, you could dig under it."

Burton Judson, the plant manager, said, "If you want to, you can build scenarios straight through 'Mission: Impossible.' Sure, you could storm this place. If twelve people drove up with guns and a truck, they could take it. Those double doors to the plutonium-storage corridor are just doors. If somebody really wanted to batter them down, they could. They could, for that matter, come in with a bazooka. But once they have the plutonium, how far are they going to get?"

I asked him what would happen if he himself were to try to steal some plutonium.

He said he would need an accomplice in the control room. It was necessary to unlock two successive doors to open a way from the plutonium corridor to the outdoors. In the space between the two doorways was a telephone. Judson said he had a key to one door but would have to call the control room to ask that the other door be opened by a button on the panel. If the man in the control room were not his accomplice, he would become suspicious when Judson, on television, started manhandling birdcages full of plutonium.

Judson, rapid of speech, candid, was a chemical engineer, still in his forties, educated at M.I.T. He used phrases like "dad-gum complicated" to describe his amazing reprocessing plant, and on its complications he held a patent. Unlike most people in the private nuclear industry, he had thought a lot about the paramilitary implications of the material he was working with. He said, "The amount of plutonium needed for a bomb is a steady figure, whereas the figure for throughput of plutonium-239 in a place like this will go up and up and up." His plant will perform a service for a fee. The nuclear material belongs to companies like Connecticut Yankee Atomic Power and Southern California Edison. So the work is done in batches. It is possible to estimate quite closely—using such known figures as average radiation exposure, reactor power levels, length of time in the core—how much plutonium is in the radioactive fuel assemblies that enter the plant. At the end of a run, the amount in the batch is measured and the small difference is noted. Someone in the plant who wanted to take plutonium home with him could probably do so without detection if he never became greedy but always operated so as not to widen alarmingly this margin of difference. It takes about twenty-five days for an average batch of fuel assemblies to be processed, and after that the plant shuts down to clean out the canyon and assay the plutonium. As the industry expands, pressure on reprocessing plants will undoubtedly grow until the batch system is too much of a luxury, and the customers will no longer be getting back their own uranium and plutonium but will be taking their share of what comes from a continuous operation that runs twenty-four hours a day. Within such a framework, it would be vastly more difficult to keep accurate books on the flowing plutonium. "It's a serious matter," Judson said. "The utilities are not interested in atom identification. They're interested in money. We are interested, though. Once you've grabbed something like this, you can't let go. You're committed to a big responsibility for a long time."

John Van Hoomissen was at the Morris plant when we were there. He was based in California and was in charge of nuclear-materials management for all of General Electric. Three people under him worked at Morris, counting atoms. Judson was not their boss. So if Judson, or someone under him, were to start siphoning off some plutonium, Van Hoomissen's men would not feel—would be less likely to feel—inhibited about reporting it. A heavyset man with an appraiser's eye, Van Hoomissen seemed to take everyone present—Judson, Fine, me, Taylor—with a grain of doubt. "All sampling here is centralized in one gallery," he said. "This safeguards against someone bleeding the sampling line. That could never happen here at Morris, but I'll show you other places where it could happen, because funny little sampling lines are run in here and there, and on a given night someone could run a funny little sampling line off to a clandestine place. The thief wouldn't have to worry much about radiation. The most vulnerable place is the nitrate point, where the plutonium comes out of the spigot. We know this. We are aware of it. A reprocessing plant used to be thought of only as the place where you got your uranium back and your plutonium credits. Now it's seen as more than that. It is not an unattended problem."

The solution, as Van Hoomissen saw it, was for the plutonium to be moved rapidly out of the reprocessing plant and back into a power reactor, where it could be burned as fuel. Plutonium is more fissionable than uranium, after all. Except in test situations, no plutonium is used in present commercial reactors, although the companies that own it have a great deal of it in reserve. The reason for this is that plutonium is one of the most toxic substances ever known in the world. Cobra venom is nowhere near as toxic as plutonium suspended in an aerosol. You could hold an ingot of plutonium next to your heart or brain, fearing no consequences. But you can't breathe it. A thousandth of a gram of plutonium taken into the lungs as invisible specks of dust will kill anyone—a death from massive fibrosis of the lungs in a matter of hours, or at most a few days. Even a millionth of a gram is likely, eventually, to cause lung or bone cancer. Plutonium that enters the bloodstream follows the path of calcium. Settling in bones, it gives off short-range alpha particles, a form of radioactivity, and these effectively destroy the ability of bone marrow to produce white blood cells. Plutonium is rendered generally in one of three forms: metal, nitrate, oxide. The oxide is a fluffy yellow-green powder. It can be fine enough to be inhaled. The oxide is the form in which plutonium would be used as reactor fuel. Therefore, it is both difficult and dangerous to make plutonium-uranium-oxide fuel pellets and slip them into zirconium-alloy fuel rods—the process necessary for use of plutonium in power reactors. Special fuel-fabricating plants would have to be built, equipped with .03-micron absolute filters, continuous air monitors, glove boxes (workers put their hands into gloves that are in effect segments of the walls of glass boxes, and handle plutonium within), and other costly equipment, nearly all of which is unnecessary in a plant that fabricates uranium fuel. So the plutonium piles up—good fuel, but uneconomical. Plutonium is worth about ten dollars a gram, and is many times as valuable as gold. As time goes by, the utilities are building up millions of dollars' worth of plutonium in their stockpiles. Meanwhile, with ever-higher extraction costs and increasing demand, the price of uranium rises. In a present-day power reactor, only three per cent of the uranium fuel is used, because the uranium-235 fissions with unprofitable efficiency after that point. After uranium itself is reprocessed, it is supposedly enriched again and then refabricated as fuel and returned to the reactors, completing a closed circuit known as the nuclear-power fuel cycle. Actually, the isotopes U-232 and U-236 present in used reactor fuel are unwelcome in the enrichment cascades. As William Higinbotham, of Brookhaven, has put it, the U-232 and U-236 would "crap up" the uranium there. So the nuclear-power fuel cycle, much advertised for its conservational appeal, is not closed, and has never been closed. The reprocessed uranium is set aside. The uranium that goes into power reactors is new uranium. The result of all this is that two economic lines are moving toward each other, and soon they will cross—a uranium line and a plutonium line—and where they will cross is the point at which it will become economically sensible to build all the necessary expensive equipment and to begin adding plutonium to the fuel that keeps present-day reactors going. This new era, which will probably arrive in full force in the late nineteen-seventies or early nineteen-eighties, is known in the business as plutonium recycle.

Recycling might open more problems than it closes. While it is an obvious way to burn up accumulating plutonium, it will also cause that plutonium to come out of storage and be circulated throughout the United States. By truck or air, it will travel to fuel-fabricating plants as nitrate or oxide, fifty or so kilograms per shipment. Then it has to be transported, often many hundreds of miles, to the reactors that will use it. About ten thousand kilograms of privately owned plutonium will be produced in 1976, fifteen thousand in 1978, twenty-five thousand in 1980. By A.D. 2000, according to A.E.C. forecasts, something over a million kilograms of plutonium will annually be travelling to two or three thousand nuclear power plants in fifty-odd countries.

It is possible to design a reactor that will produce

more plutonium than it uses. The Atomic Energy Commission has published a four-color poster advertising this. The poster says: "Johnny had 3 truckloads of plutonium. He used 3 of them to light New York for 1 year. How much plutonium did Johnny have left? Answer: 4 truckloads." In this way, the A.E.C. has been introducing the public to breeder reactors. They are not new. The first power reactor that ever lighted a bulb was a small, experimental breeder in Idaho. A sixty-megawatt demonstration breeder reactor was built by Detroit Edison on the shore of Lake Erie, but it was beset with operating problems and eventually shut down. The idea of the breeder is to use a combination of fissionable and fertile material, making heat with the one and new fuel with the other. The fissionable material, for example, might be plutonium-239 and the fertile material uranium-238—ordinary, natural uranium. As the plutonium fissions, it throws off many more neutrons than are needed to keep the plutonium chain reaction going. The excess neutrons go into the nuclei of the U-238, which becomes U-239, which decays to become neptunium-239, which decays to become plutonium-239, ready now to get into the original chain reaction, ready to repeat the process and produce even more plutonium. Because the fissioning plutonium puts out many extra neutrons and because there is a high proportion of fertile U-238 in the reactor core, the breeder makes more plutonium than it uses up. Theoretically, the breeder can make more than fifty times better use of uranium than present-day reactors. Moreover, it could use as fertile material the two hundred thousand tons or so of leftover U-238 that has been separated from U-235 since the military weapons program began. Breeders are variously cooled by salt, sodium, helium; and they have a fine set of names: the Molten-Salt Breeder Reactor, the Liquid-Metal Fast Breeder Reactor, the Gas-Cooled Fast Breeder Reactor. The Germans have one called SNEAK. The French have one called Rapsodie. They are research reactors. In 1973, the Soviet Union announced that it had begun commercial power production with a breeder at Shevchenko, on the Caspian Sea. Breeders as a working generation are still some time away, but when their time comes the figures for world flow of plutonium will be not so much increased as multiplied. So will the probabilities of the clandestine manufacture of atomic bombs.

Where is plutonium now—that is, plutonium owned by private companies? In greatly varying amounts, it is in Hanford, Washington; West Valley, New York; Pawling, New York; Morris, Illinois; Erwin, Tennessee; Pleasanton, California; Crescent, Oklahoma; Cheswick, Pennsylvania; Leechburg, Pennsylvania; and in transit among these places. It has ridden around the country sometimes with ordinary truck freight—linoleum, Congoleum, plutonium. New regulations forbid this. A ten-litre bottle of plutonium-nitrate solution in a birdcage—two and a half kilograms of plutonium—was shipped from Hanford to Crescent not long ago at the rear of a flatbed truck. Other cargo filled up the bed space, and the plutonium, the last thing on, was held by a single chain. It was clearly labelled "DANGER—PLUTONIUM." Generally, the material goes by itself, in shipments of about fifty kilograms. Plutonium-uranium fuel pellets are made at Crescent by Kerr-McGee, and are put inside metal rods and sent back to Hanford, to the A.E.C.'s Fast Flux Test Facility—an experimental breeder reactor. Kerr-McGee handles about a thousand kilograms a year. So does NUMEC (Nuclear Materials and Equipment Corporation), in Leechburg, which also makes fuel for the Hanford facility. Fuel rods for plutonium recycle are being made by General Electric at Pleasanton, by United Nuclear in Pawling, by Nuclear Fuel Services in Erwin, and by Westinghouse in Cheswick.

The Cheswick plant makes breeder fuel as well. Up to a hundred kilograms of plutonium may be there at any one time. Nine security police work every shift. They are armed and have been instructed in the use of weapons. An eight-foot fence surrounds the building. In one corner of the plutonium-oxide laboratory are forty safes; each safe is designed to contain two kilograms of plutonium in a can. If the continuous air monitors sound their alarm—oxide in the air—the entire building can be evacuated in sixty seconds. There are many doors, many of them coded red on one side. This is a pilot plant. It indicates the conditions under which the great abundance of future plutonium will have to be handled. Rooms are filled with assembly lines of glove boxes. Some contain tiny pellets of plutonium-uranium fuel for the Fast Flux Test Facility. A half inch long, they look like horse feed. Others hold plain plutonium oxide. It has the consistency of flour. Dust. Yellow-green dust. Because of its fine consistency, it has a peculiar locomotion. Spontaneously, it creeps. It moves around. It spreads like lampblack. Drop a little of that in an air-conditioning system and a whole company will die. A man who works in Cheswick says, "We're picking up a lot from the baby-food industry. They keep people away from the food. Here we keep the plutonium away from the people." Looking through the glass walls of a glove box, another man says, "It would require a fair amount of skill to get that stuff out of here without crapping up the countryside."

In 1971, the Kansai Electric Power Company removed some fuel assemblies from its Mihama No. 1 reactor, in Fukui, Honshu, Japan. The radioactive fuel rods contained fifty kilograms of plutonium. In heavy casks, the fuel assemblies were shipped to England. They went to Windscale—a reprocessing plant in Cumberland. Later that year, the fifty kilograms of separated plutonium, in oxide form, was shipped by BOAC to Kennedy International Airport. A courier rode along. At Kennedy, the material was met by a man from Westinghouse and was loaded onto a truck (Forest Hills Transfer) that carried no other cargo. It was driven, on the New Jersey and Pennsylvania Turnpikes, to Cheswick, which is a few miles east of Pittsburgh. In Cheswick, the fifty kilograms of Japanese plutonium were fashioned into pellets of mixed plutonium and uranium oxides, and placed inside seven hundred and fifty zirconium-alloy rods. After the rods had been fitted into assemblies at another plant, in Columbia, South Carolina, they would be shipped back to Japan. Plutonium recycle would then begin in the Mihama reactor. Fifty kilograms is almost ten times the amount that was used in the Nagasaki bomb.

West Valley, New York, is a small town in Appalachian-foothill terrain, about forty miles east of Lake Erie. A sign tells approaching drivers, "SEEK YE THE LORD WHILE HE MAY BE FOUND." A sign tells departing drivers, "BELIEVE IN THE LORD JESUS CHRIST AND THOU SHALT BE SAVED." There is a blinking red overhead light, a chain-saw shop, a visiting bookmobile, and an old wooden hotel with rubber-booted clientele and both Schmidt's and Schlitz on tap. Also in West Valley, there is enough plutonium to arm a nation.

For many years, West Valley was the only place in the country where fuel from commercial reactors was reprocessed. The plant there, which belongs to Nuclear Fuel Services, is much like the one at Morris, Illinois, with its canyon, its load-out cells, its plutonium-storage room—eighteen by thirty-six feet, sixty-eight birdcages maximum, a door with padlock and chain. I went there with Ted Taylor one day. "We're sort of proud of our pickle works," the plant manager told us. "We're pioneers." He noticed that my hand was resting on the rim of an empty plutonium birdcage. "I wouldn't touch anything," he said. "A little of that goes a long way." His name was James Duckworth, and he was a chemical engineer who had worked for fourteen years at Hanford before coming to West Valley, in 1967. A thoughtful, practical, kindly person, he was worried about the international aspects of safeguarding so-called special nuclear materials (weapons-grade materials). He was worried about his sons at Syracuse and Cornell. He was worried about the great vortex of changes in the society. "We of the Establishment resist change," he said. "But we do the very things that advance it." Taylor asked him if he ever worried about the possibility that plutonium might be stolen from his plant.

"No. Honest to Pete, no," he said. "I have so God-damned many real problems. I haven't time to imagine them." He agreed that a pickup truck containing two people and two guns would constitute a force sufficient to remove from the plant as much plutonium as the truck could carry—that is, about nine birdcages, or twenty-two and a half kilograms.

In 1967, in a Nick Carter paperback called *The Weapon of Night,* nine Chinese went to Duckworth's plant and stole some plutonium. Their adventure ended in failure underneath Niagara Falls. Those Chinese went to the right place at the wrong time. They should have waited a few years. In 1967, when the government was buying plutonium from the private companies, plutonium nitrate was regularly shipped across the United States from West Valley to the A.E.C.'s storage tanks at Hanford. After 1970, when "plutonium buyback" came to an end, various utilities, like Consolidated Edison and Pacific Gas & Electric, began wondering where they could stockpile their plutonium. They turned for help to the New York State Atomic and Space Development Authority, which owns the land on which the West Valley reprocessing plant was built. The Authority built a warehouse. It stands alone in a clearing in woodland, something over a mile cross-country from the reprocessing plant. No other buildings are close to the warehouse, or even visible from it. A broad, paved drive called Buttermilk Road—followed by an overhead power line—runs a half mile in from New York 240, a blacktopped country road. The warehouse is one of a kind. In purpose, there is nothing like it anywhere. Known as the ASDA Plutonium Storage Facility, it is made of steel panels painted pastel green. One man works there eight hours a day five days a week; otherwise no one is there. Its dimensions are eighty by a hundred and sixty-three feet. It has one window and three doors. It is surrounded by a seven-foot chain-link fence strung along the top with barbed wire. Its other protective devices, which are extensive, are not apparent to the outside observer. Its design capacity is two thousand kilograms of plutonium.

The nearest neighbor of the ASDA storage facility was an old general store on New York 240, with a row of decaying tourist cabins in back, a couple of gasoline pumps in front, and an outdoor clock whose hands did not move. I went into the store, bought a Hershey bar, and asked the storekeeper what that green steel building was on Buttermilk Road. He was a white-haired man wearing rubber boots and a red checked shirt. "That? That's where they keep the potent stuff," he said.

In a chair opposite the candy counter sat another white-haired man wearing rubber boots and a red checked shirt. He said, "Yes, sir. That's where they keep the potent stuff."

A sign on the wall said, "Cows may come and cows may go, but the bull in this place goes on forever."

THE United States Navy, in 1945, did not choose to exploit Ensign Taylor as a physicist. His request for a billet in atomic research was overlooked. Instead, he was sent out to the Pacific on an attack-transport to collect and bring home personnel from outposts of the war. One such place was Eniwetok Atoll, in Micronesia. When the ship was loaded and was steaming away, he stood by the rail, looked back at the receding islands, and thought, I don't know where my life will ever take me, but one thing certain is that I will never see this place again.

The Navy let him out in the summer of 1946, and he enrolled in graduate school at the University of California at Berkeley. To him, the academic world seemed almost purposefully designed to make subjects uninteresting. At least, this had been true at Cal Tech, where the art of teaching had seemed to consist entirely in pointing out errors, and now, at Berkeley, with only an exception or two, nothing much happened to change Ted's basic view. Most courses were lecture courses. Most lecturers were functional gargoyles pouring forth unrelated facts. Ted chose not to listen. He did study. He studied hard, following his own interests without much regard for the broad and general picture. He was bored by some quite basic subjects. "Thermodynamics was dull. Sometimes I think I am incapable of understanding something I am not interested in. I studied it but did not learn. A lot of physics was a mystery to me, and still is."

He went down to Claremont whenever he could, to Scripps College, to see a girl he knew. Her name was Caro Arnim, and she was majoring in Greek. She was writing her thesis on the *Electras* of Sophocles and Euripides. She was athletic, dark-haired, blue-eyed. She wore glasses. If anything, she was even more shy than he was. She spoke in a voice so soft that it could almost disguise the acuity of what she had to say. She would someday be a librarian. They got along by whistling to each other. They whistled the themes of classical music. See if you know what this is. A kind of test. "Ted was really good at that sort of thing. I was surprised. We whistled themes from Beethoven symphonies, from Handel, from Bach. We tried to remember them. For people who are shy, that's not a bad way to start, if you have trouble talking."

She found him attractive—tall, gangling, with a broad forehead, a somewhat pointed chin, and great thoughtful brown eyes, which often seemed to be focussed on something no one else could see. "When we got married, he was going to be a college professor in a sleepy town. In those days, both of us were unsure. We were about the shyest people you ever met in the world. How we had the courage to talk to each other seemed a wonder sometimes. A sleepy college town was about our speed." They went to the beach, sat on a sand dune, and talked immortality. Within his enthusiasms, he could persuade her of almost anything, but with immortality she was somewhat bored. Ted took some getting used to. In their apartment in Berkeley, he would sit and look straight at a wall for vast tracts of time. She feared that there was something wrong, and that she might be at fault; but he was simply thinking. Sometimes, she tried whistling. "Do you know what that is?"

"Oh . . . That? *Variations on a Theme of Frederick the Great.*"

(The Taylors now live part of the year in a house on top of a mountain in western New York. The record collection there contains a great deal of Bach, and one winter day Ted was listening to *Variations on a Theme of Frederick the Great* while attempting to prepare himself a cup of instant coffee. A kettle was beginning to steam. He put some powdered coffee in a cup. He looked out a window into slowly falling snow. "Such a simple theme," he said. "The variations must have been the product of a very clear thinker, because the patterns are such a systematic exploration of a lot of different possibilities. Up pyramids. Down pyramids. There's a periodicity to it. Structural patterns like those are the kinds of things that appeal to a theoretical physicist—the combination of predictability and surprises. The measure of greatness of a composer is his ability to combine these. The way I like to think about physics is that there is an exact analogue to the composer, the creator—the knack that Bach had for putting the world together in a way that is somewhat predictable but also full of surprises. One of the reasons that Bach's music is so satisfying in this respect is that he was a very religious man, and I suspect that he was getting some instructions in how to do this by simply listening to his Maker." Steam was pouring from the spout of the kettle, but Ted had become so absorbed in what he was saying that he reached over and made his coffee with warm water from the kitchen tap.)

One of the people who taught him physics at Berkeley was J. Robert Oppenheimer. Ted found him "a good teacher for bright students." Since Ted was not a bright student, he did not experience Oppenheimer's talent. With several other students, he once went to Oppenheimer with a written proposal for a general strike of all physicists in the United States. Oppenheimer said, "Take this paper. Burn it. Never recall it. Anyone who knew of this would label you a Communist and you would have no end of trouble the rest of your life." Ted worked part time at Berkeley's Radiation Laboratory, mainly on the cyclotron, also on a beta-ray spectrograph, for which, with other students, he designed some novel features. Noticing this effort, Ernest O. Lawrence, the laboratory director, decided that grad students should not be doing such work. The physicist Luis Alvarez said to Lawrence, "These young men are going to go very far." Not on this project, they aren't, said Lawrence, in effect, and a senior man finished what the students had begun. It had been quite a battle—Lawrence versus his subordinate colleagues. Ted, just a student, was bitter. He thought, If that is what experimental physics is, to hell with it. So he went into theoretical physics, and found it to be much more his natural milieu. He worked under Robert Serber, who had been an instrumental figure in the Manhattan Project, had helped construct the mathematical framework of the first bombs, and had written a compendium of the physics of atomic bombs—a work that was called *The Los Alamos Primer.* Ted took Serber's course in neutron-diffusion theory, and he planned a doctoral thesis predicting the characteristics of the scattering and absorption of neutrons by nuclei. Oral preliminary examinations came along. He took one on mechanics and heat. The examiners—three senior professors—watched while he tried to derive on a blackboard a formula having to do with the Second Law of Thermodynamics. He became confused and rattled. He simply did not know enough to do what he was being asked to do. He flunked. He took a second prelim, this one in modern physics, and he failed again. Low on self-discipline, high on enthusiasm, he had followed his interests ("I liked to do what I liked to do"), and had not spread himself sufficiently over the fields he was supposed to know. Moreover, he was nervous. He was numbed by the examination procedure, had always been afraid of exams. "We cannot in good conscience pass you," he was told. "We realize this is the second time. You can't remain here at Berkeley as a graduate student."

Serber, though, thought of his student as a person of special and unusual ability; he pondered the loss of him to the world of physics, and he hoped that would not happen. There was no university worth the name that would welcome Ted at this particular moment in his academic performance. Where could he go? He was not a scholar, not a profound and thorough analyst. It did not take much perception to see that. He was more a conceiver of things. Serber picked up a telephone and called J. Carson Mark, director of the Theoretical Division at Los Alamos Scientific Laboratory. Serber told Ted what nice country there was around Los Alamos—perfect for hikes; he would like it there. Mark had agreed to give Ted a try in New Mexico. Ted was grateful. His confidence was way down—as low, probably, as it would ever be in his lifetime. As he and Caro packed and left Berkeley, he did not even have a clear idea what he would be doing at Los Alamos. His work, as he understood it, was to be "in neutron-diffusion theory." Twenty-four years old—it was November, 1949—he was a little taken aback when, soon after he was shown to his desk at Los Alamos, he was handed drawings of uranium and plutonium bombs.

The Los Alamos Primer, which contains the mathematical fundamentals of fission bombs, was declassified in 1964 and is now available from the Atomic Energy Commission for two dollars and six cents a copy. For four dollars, a book titled *Manhattan District History, Project Y, the Los Alamos Project* can be bought from the Office of Technical Services of the United States Department of Commerce. Written in 1946 and 1947, this was the supersecret technical description of the problems that came up during the building of the first atomic bombs. The book was declassified in 1961. On its inside front cover is a legal notice that says, in part, "Neither the United States, nor the Commission, nor any person acting on behalf of the Commission . . . assumes any liabilities with respect to the use of, or for damages resulting from the use of, any information, apparatus, method, or process disclosed in this report." Long and various is the bibliography of works in public print that contain information that was once of the highest order of secrecy. The release of documents containing detailed information on the sizes, shapes, design, and construction of nuclear explosives—and on such topics as plutonium metallurgy and the chemistry of initiators—seemed to follow, over the years, a pattern of awareness of Russian knowledge. When it became clear that the Russians knew about something or other, what then was the point of keeping it secret?

The Atomic Energy Act of 1946 included almost no role for private persons. In 1953, President Eisenhower introduced his program called Atoms for Peace, its idea being to share the atom with all the world for the benefit of mankind, for the development of emerging nations, for the making of what was described as "meterless power." Much debate followed. Critics of the program called it Kilowatts for Hottentots and pointed out that a reactor exported to the African bush would not be particularly useful there unless a staggering amount of additional capital was exported with it. Others suggested that the United States, as the nation that had opened the nuclear era with more than a hundred thousand deaths from nuclear bombs, was now attempting to sublimate its guilt with a program that was styled to do good but could in the end bring evil, for every new reactor would be making plutonium, and plutonium atoms multiplying everywhere were hardly a guarantee of peace. After the debate, the Atomic Energy Act was amended, in 1954, and the way was now open not only to reactors in Bechuanaland but, on a much larger scale, to reactors in New York, New Jersey, Illinois, California. As it happened, though, the electric companies were quite reluctant to go nuclear. Arithmetic revealed that "meterless" power would cost more than the kind the companies were already selling. The long-range safety of nuclear power plants was unknown. No insurance company would write a policy to cover a reactor. This put the Atomic Energy Commission in an interesting dual position. An energy crisis was obviously coming. A kind of mandate had been issued with the promulgation of Atoms for Peace. So it fell to the A.E.C., the agency that had been established to control atomic energy, to promote it as well. The apparent conflict of interest was quickly mitigated, though, because, as the agency rapidly expanded, it expanded principally in the direction of promotion. The United States government offered to become the insurer of power plants. The United States government would help build demonstration reactors. The United

States government would lease slightly enriched uranium at an attractive price and buy back unwanted plutonium at an even more attractive price. Moreover, if the utilities really did not want to go nuclear, possibly the U.S. government would go utility. The electric companies went nuclear. Soon thereafter, the Atomic Energy Commission moved out of downtown Washington to rural headquarters near Germantown, Maryland —a marathon from the center of the city.

While most people within the A.E.C. were concerned with the development of the nuclear industry, some worried about the implications of proliferating plutonium. Finally, in the middle nineteen-sixties, a panel was set up to look into the matter, and two new divisions were established to deal with safeguards. Licenses that had been issued by the A.E.C. to private companies were renegotiated in order to spell out in a formal way requirements for safeguarding nuclear materials.

A semantic distinction developed between safeguards and safety. Safety meant environmentalists snapping about emergency core-cooling systems, thermal fishkills, radiation clouds, "the China syndrome" (the reactor melts and starts down through the earth for China). Safeguards meant keeping track of and protecting the materials that could be turned into bombs. It meant vaults, alarms, fences, locks, guards, and German shepherds. And it meant accounting. By both chemical and nuclear means, it was becoming ever more possible to count atoms and, more or less, to balance books—to sense if material had been stolen or embezzled, and, with approximate accuracy, how much. People are now working on various aspects of safeguards at Germantown, M.I.T., Brookhaven, Los Alamos, and elsewhere. They develop, for example, instruments that can read the characteristics of fissile material—gamma emissivity, alpha emissivity, isotopic abundances (how much U-235? how much U-238?)—without having to destroy what contains it (fuel rods, waste-storage drums, or whatever). The idea is to enable a company to count its uranium or its plutonium and be able to say how much, after a given process, may be missing. An analogy is often drawn between special-nuclear-material balances and bank balances, but the analogy is imperfect. It is impossible to balance the books on nuclear material. While being machined or sintered or compacted into pellets, some inevitably gets lost. This is known as MUF—Materials Unaccounted For. The cumulative MUF at a large fuel-fabricating plant can amount to dozens of kilograms a year. The MUF problem cannot be eliminated, but it can be minimized, which is what safeguards specialists are attempting to do.

Safeguards, ideally, are a series of frames around the nuclear industry, expanding with it through time. As the industry multiplies—as plutonium recycle comes in, and the breeders, and the fully enriched uranium of the H.T.G.R.s—ideal safeguards systems would pace the industry, a little ahead of it. These would be commensurate safeguards, and one would imagine that nothing less would do. But there are problems. To start with, it is not possible to say with precision what commensurate safeguards should be—not in general, and not even in a particular situation. Two people of equal training will judge differently what is adequate. One extreme recommendation is that the nuclear industry, as a source of bomb material, is too dangerous and should be shut down. The opposite extreme is to decide that no person or group would ever steal special nuclear material, and even if material were stolen it would require another Manhattan Project to produce a bomb, so such worry is groundless and safeguards are unnecessary. Dozens of slightly varying positions are taken along the bridge between these points of view. Subjective influences are obviously present. Someone who has spent a fair part of a career perfecting a Lithium-Drifted Germanium Gamma-Ray Spectrometer obviously believes in the need for such a machine in accounting for nuclear material. Someone who has given years to the development of the High-Temperature Gas-Cooled Reactor would not tend to begin a discussion of its features by pointing out that it uses weapons-grade fuel. If one can imagine commensurate safeguards, one can also imagine veneer safeguards—and different people might use the one word or the other to describe the same situation. As new A.E.C. safeguards requirements have come along—from early ones in 1967 to the more recent ones in 1973—reactions have ranged from the complaints of industry (too much interference) to the dismay of people like Ted Taylor, who feel that the requirements are still little more than veneer, inadequate for the present, let alone the future. Difficulties mount. Safeguards cost money, which means a diminution of profit, putting kilowatts out of reach. Safeguards suggest dangers, which belie the promise of Atoms for Peace, and thus can be a hindrance to the promotion of nuclear commerce. The Atomic Energy Commission, as a whole, is profoundly dedicated to the growth and spread of nuclear power, and I have heard one of its commissioners (James Ramey), in addressing a large audience, say, "We in the atomic-energy industry . . ." Safeguards are inconvenient to an industry that does not want to frighten its corporate or individual customers, suggesting war instead of peace. Indications are that in the Soviet Union no nuclear material of any kind travels anywhere except under convoy by the Red Army. A suggestion that the United States Army be used in the same way was rejected because such military involvement would create a bad image for the industry. In the annual struggle for budget, people who concentrate on safeguards have to appeal for money, like everyone else. Since numerically they are less than two per cent, they constitute a voice that is somewhat muffled within the bureaucracy. I once asked Delmar Crowson, a retired Air Force general who was the A.E.C.'s director of Nuclear Materials Security, how difficult it was for him to implement new safeguards that he might consider essential. He shrugged, smiled at some colleagues, and said, "There's the problem."

Russell Wischow is a nuclear C.P.A., more or less. He is the president of the Nuclear Audit & Testing Company, which is based in Washington. His approach to the safeguards problem is, as he puts it, pragmatic, and he thinks that for many people it is, unfortunately, an emotional issue. He thinks the A.E.C. is too often "considered guilty until proved innocent." He thinks the private nuclear industry is here to stay, that there can be no reversion to a government monopoly, and that industry—as a responsible unit of society, regulated by the A.E.C.—will do the best it can. "What else can you do? Put the material in a vault and turn the vault back to 1942?"

I had called on Wischow in a suite his firm had taken at the Hotel Shoreham in Washington during a meeting of the Atomic Industrial Forum and the American Nuclear Society. A tall, elegant, dark-haired man in his forties, in buckle shoes, a black suit, a shirt in stripes of pink and gray, he had once taught in the reactor school at Oak Ridge and had worked some years in West Valley. He spoke informally over bits of pineapple wrapped in bacon. These were some of the things he said:

"It's not like a bank. You cannot balance the books. What's wrong with a MUF of a few dozen kilograms of plutonium if your throughput is such that you can't measure it closer than that?

"I guess this is a dangerous statement, but I'm going to make it anyhow. If there were real intent to divert material, you could get away with it. You can't be greedy. You have to work within the limits of measurement.

"What are we trying to do—keep a bomb out of the hands of a country or a few grams out of the hands of a group? If you want a few grams of plutonium, you can steal that almost anyplace in the country. The safeguards problem is out of focus. No one is willing to state how much.

"I can't believe that a company would divert. Individuals, yes. If any segment of the industry wanted to divert, it could—gram quantities, kilogram quantities. When you found it out, it would be too late.

"Guards in some places have guns but no bullets. No company pays its guards enough to ask them to throw down their lives for material.

"Somewhere between the intensity of Ted Taylor and the lackadaisical attitude of some in the industry is reality.

"Safeguards are frustrating. The stuff is difficult to quantify. You can't put it into a vault and keep it there. Once it is in the manufacturing cycle, you open it up to pilferage. I'm very concerned about the end results of a safeguards system that doesn't work."

Before forming the Nuclear Audit & Testing Company, Wischow was director of Nuclear Materials Safeguards at the A.E.C. He was replaced by Charles Thornton, whose experience went back to the Manhattan Project. Thornton helped set up labs at Oak Ridge in 1943, and worked on isotopic separation there. I sought him out at the same convention in Washington. A small, thin, wiry man with white hair to his shoulders and rimless glasses, he looked to me remarkably like Benjamin Franklin, a condensed Benjamin Franklin, although he was wearing a plaid suit, a checkerboard shirt, a wide gold tie, and a snakeskin belt. These were some of the things Thornton said:

"All the guys who tell you that American industry is experienced in protecting its vital materials—that's a crock. Mankind has never handled as dangerous a commodity as plutonium. We have never developed the skill.

"Plutonium is worse in its toxicity than as a bomb. Plutonium is worth, at most, ten dollars a gram. If the Black September organization had a hundred grams of this material they could wreak havoc. The Fast Flux Test Facility will have two contracts handling seven hundred kilograms of plutonium each. That is more than ten million dollars' worth of material. A thousand-dollar loss is thus insignificant.

"One gram equals one times ten to the sixth micrograms—a million micrograms—and if it were properly distributed it would bring one times ten to the sixth fatalities. A microgram inhaled can cause bone cancer. Take what people think they're worth in terms of dead. At least twenty-five thousand dollars, right? One times ten to the sixth times twenty-five thousand dollars is twenty-five billion dollars. Such criteria might be used to determine the intensity of the constraints put on the industry.

"The aggregate MUF from the three diffusion plants alone is expressible in tons. No one knows where it is. None of it may have been stolen, but the balances don't close. You could divert from any plant in the world, in substantial amounts, and never be detected. In a diffusion plant, take any pipe and freeze out the material that is passing through. Set up a diversionary pipe. Cool it with liquid air, and get it into a bottle. Coca-Cola trucks go in and out of the restricted areas there all the time. All sorts of people.

"The statistical thief learns the sensitivity of the system and operates within it and is never detected. Scenarios to get stuff out of the cascades are as varied as the ingenuity of individuals. Put a saddle valve into a pipe. Cool the pipe with methyl chloride. Take a saltshaker each day in your lunch bucket. Take a hundred grams a day. A kilogram every ten days. Or hit the shipping point. Or doctor the record of sampling bottles. That would not be my choice, though. Fully enriched uranium in a conversion plant—a pale-yellow fluid—could be put in a hot-water bottle under your shirt.

"The A.E.C. can say officially that quantities of MUF are not dangerous. This is not so. Tons have been lost. They can say they have impregnable barriers, sensitive modern instruments. Not that impregnable, not that sensitive. They can say, 'The numbers are not good, but we don't know how to do any better.' If you admit that this industry is not controllable, then you shut down. You wait until it is controllable, and then start up.

"The incremental capital for an adequate safeguards system would not destroy the industry—if it were designed in.

"It's late in the history of the world to go into the safeguards business."

Thornton—whom William Higinbotham, of Brookhaven, has described as "a great undiplomatic breath of fresh air who rattled everybody"—lasted a year and a half in the safeguards job, and then, by his own description, was "fired." His detractors called him a pedant. He was not actually fired. He was lateralled off the field. Remaining in the employ of the A.E.C., he became Special Assistant for Energy Policy, Office of Planning and Analysis.

India, Italy, and Japan have reprocessing plants capable of removing plutonium from spent reactor fuel on a laboratory scale. Big power reactors are on line and making plutonium in India, Pakistan, East and West Germany, Japan, Spain. Only several thousand kilograms of weapons-grade nuclear material exists now outside the five nations that have exploded bombs. The figure is steadily growing. Safeguarding special nuclear material is basically an international matter. Some thirty nations will have reactors and will be producing weapons-grade material by 1980. Furthermore, material stolen from one country could be used by a second country against a third. Because only a small quantity of material is needed to do immense damage, international safeguards are analogous to a simple chain, and until it is too late to do much in a preventive way it may be impossible to tell which of many links is the weakest. Responsibility for international safeguarding lies with the International Atomic Energy Agency, which has its headquarters in Vienna and sends inspectors—nuclear auditors—to nations that have agreed to cooperate, either under the terms of the nuclear-nonproliferation treaty of 1968 or by some earlier arrangement already existing between a given nation and the I.A.E.A., which was established in 1957. Not all nations cooperate.

Henry D. Smyth, professor emeritus of physics at Princeton, was United States Ambassador to the International Atomic Energy Agency from 1961 until 1970. He has often been called a nuclear statesman. He worked on the Manhattan Project—among other places, in the Metallurgical Laboratory in Chicago—and it was he who was chosen to write a book to be published immediately after the war explaining publicly what had happened, what had led to the new phenomenon. From 1943 onward, his office on the Princeton campus had armed guards outside the door round the clock, because Smyth, some of the time, was in there describing the development of a type of weapon that would end the war, no matter who exploded it. His book, *Atomic Energy for Military Purposes,* remarkable for its concision and its lucidity, was published in 1945. It contained the basic physics but not what were then the secrets of the fabrication of the bombs. From 1949 to 1954, some years before he began to commute between

Princeton and the I.A.E.A. in Vienna, he was one of the five commissioners of the United States Atomic Energy Commission. I went to see him one day not long ago at his office in Princeton. He is a tall and angular man with steel-gray hair, mildly formal in manner but without starch. I asked him if he found it possible to be optimistic that there could be effective international safeguards.

"Yes," he said. "Let me explain. Anything is better than nothing.

"The safeguards aspects of the nonproliferation treaty were drawn up by a special committee, involving thirty or forty nations, that met on and off for a year. That such a committee could work for a year on a difficult technical and political problem and come out with a reasonable answer is in itself something of a triumph. I was surprised at the degree to which you could get cooperation from people, and the degree to which they developed national pride through being internationally minded. I think there are going to be a lot of people who are reluctant to break the club rules.

"International safeguards depend on national systems. A truly international safeguards system would be impossibly expensive. Nations would not go along with it anyway.

"Our most serious problem with regard to the nonproliferation treaty is that it emphasizes national safeguards systems—and if the United States is interested in the nonproliferation treaty, the United States safeguards system should be good and effective and orderly, and, as far as I can make out, it isn't.

"What I am concerned about internationally is power reactors in countries that have unstable governments. The reactor, wherever it is, builds up a stockpile of plutonium. Now suppose there's a revolution. A totally new and crazy government comes in, and there's the plutonium just sitting there asking to be made into a bomb.

"The A.E.C. production and reactor people couldn't care less about these international problems.

"The A.E.C. approach is 'Papa knows best. Papa is guarding against every possible danger.' You look into it and find they are not.

"I think security in this country is important, particularly protection of weapons themselves, but I think illicit production of weapons is more likely to come as a national enterprise than as the enterprise of a gang."

General Crowson agreed with that last point. "It requires a plot," he said one day in Germantown. "To get all the people together without the plot leaking seems all but impossible. A dozen, maybe two dozen, would be needed—all highly trained individuals. One man could not do it. The scenario of the home bombmaker is overplayed. That piece has been highly overplayed. Suppose you have a set of plans for a gasoline engine. How many people do you know who could make one?"

When Ted Taylor first approached Los Alamos, in

1949, he climbed into the mountains from Santa Fe in his 1942 green Buick coupé. Caro steadied a basket that held a baby. In back, where a seat had once been, was a large part of their earthly goods. Los Alamos, seven thousand feet up into the ponderosas, and not far from the Continental Divide, is built on mesas that project from mountainsides. The laboratory is twice as large now as it was when Taylor first saw it. On that day in 1949, a big Army tank was beside the gate, its cannon pointing down the road at incoming cars. Ted stopped at the gate—a guard tower to his left, a building to his right full of guards and files on all personnel, even babies. The Taylors were identified and given badges, and they went on through. Today, cars come and go. The tank is gone. The tower is empty. The guard building is a Mexican restaurant called Philomena's. The road is open.

J. CARSON MARK, the man who had agreed to have Taylor come work at Los Alamos, had been there since wartime, when he led the diffusion-theory group of Project Y, the code name of the Los Alamos project. A Canadian, as precise in his diction as in his physics, Mark was a subtle man—large of frame, a large head, a somewhat judicial demeanor. He had eventually become chief of the Theoretical Division, which now, in 1949, was about to enter its second era of extraordinary conceptual advance, its efforts directed mainly toward the perhaps insuperable problem of igniting a thermonuclear explosion. A quarter century later, Mark would still be at Los Alamos, still running the Theoretical Division, with an old Santa Fe Railroad clock on his wall; and he would look back with fondness to the days at the beginning of the nineteen-fifties when a group of young men came in from various universities to help conceive new bombs. "This was a tremendous group of people," Mark would remember. "A constellation. Bob Thorn, Walter Goad, George Bell, Ted Taylor —all new to professional work, all enthusiastically collaborating. Bob Serber, at Berkeley, had known Ted well enough to be able to assure me that we would be quite well served if he came here. Serber said that Ted had gone to pieces in his oral examination. That was not a worry to me. While we had lots of things at Los Alamos that people could get anxious about, we did not have anything quite as crucial as an oral examination. Ted was a delightful, bright young man. He was not the best from the point of view of command-depth assurance in physics, but he was far above average; and what was really outstanding was his prying into corners, turning over stones—his enthusiasm, his eagerness, his curiosity, his restlessness. These things, combined with a very *good* level of physics, made something quite unusual."

On December 10, 1949, Ted and Caro began what proved to be a long letter, written over several days, to Ted's parents in Mexico. ("Working on the bomb was a difficult thing to write home about," Caro would say in later years. "After all, Ted came from a family of missionaries and ministers. All through our time at Los Alamos, Ted's mother made guarded statements about what he was doing. Ted had talked peace in college. It was a surprise to find him working on the bomb.") In the letter, Caro wrote:

> I'm getting more used to the looks and atmosphere of the town. At first it was a shock to see so many like houses set row on row. Some of the newer, postwar places are very nice looking, though of necessity they are all rectangular or square boxes. Some of the earliest places are like slums, with little or no lawn, and a general downtrodden look. Most of these are "sub-standard" and to be torn down. Ours is sort of in the middle. I guess it's not a slum, and is very clean and adequate inside, though built with no imagination, and painted a horrid mustard and maroon on the outside. There is a patch of lawn in front which needs encouragement, and a hundred square yards of picket-fence-enclosed bare dirt in back, for laundry and mudpies. . . .
>
> The nature of Ted's work and its secrecy are sort of a family ghost, and a hindrance to companionship. Perhaps I'm too curious, or too dependent on his interests for mine, but it used to be fun to at least hear what he was doing.

Ted continued:

My work has been most interesting, and I've been learning a great deal. No doubt you remember how I felt, a couple of years ago, about ever working on anything directly connected with military applications of atomic energy. Since then my ideas of this have changed. I'm not certain when this change came; it's been slow. (I do know that it was *before* I was offered the job at Los Alamos.) I've always thought that the very existence of a means by which men could conceivably completely destroy each other might be just the thing which would prevent any future world wars—and yet at one time I claimed that I would have nothing to do with development of the bomb. This now seems inconsistent. The way I feel now is this: A full-scale war between Russia and the U.S., in which A-bombs in their present form were used, would make this world unliveable, as far as I am concerned. And yet people in Congress (and, I suppose, in the Kremlin) talk about a future war as something indeed horrible—*but* they talk in terms of preparing to win it. I claim that these people don't fully realize the destructive potentialities of atomic energy. The Bikini tests have been played down; Hiroshima is pointed at as the proof that a modern city can survive an atomic blast; comparison is made with the strategic bombing of Germany during the last war, and the conclusion is reached that atomic bombing would be little more decisive. This, I think, is all wishful thinking. It ignores the tens of thousands killed outright at Hiroshima, the effect on a nation of the destruction by conventional bombs in Germany *if* it had all been accomplished in one day, the fact that there *was* an immense amount of destruction at Bikini.

I think that there is only one realistic way to avoid war, and that is to make the world really afraid of it. I think the world should be afraid of it now, but apparently wishful thinking and ignorance (particularly on the part of those people who have some say in what goes on) have removed much of this fear. If A-bombs in their *present* form will make another war something which mankind cannot bear, and if most people don't realize this, then, I say, there is only one thing to do: develop a bomb which will leave no doubt in *anyone's* mind. This idea is repulsive to most people I know, and yet I feel, as strongly as I have ever felt anything, that it is the only way out. The basic physical principles of a superbomb are all there. If a war with conventional weapons did not effectively wipe out civilization (as I think it would), I am certain that a superbomb would be developed during the war, as it was during the last, and would be used until civilization really was wiped out. So, again, I think that the thing to do is to find that horrible thing *now*, before a shooting war starts and people completely lose their ability to reason. Once fear of war removes the immediate threat, then my idealism takes over, and I think in terms of World Government.

Enough of all this, for now. I can just say that I firmly believe that what I am doing now is right, and that I will continue to do it until someone or something shows me a way in which I, personally, can do more to help prevent another war, or that I am wrong.

Lots of love,
TED

When he spoke of a superbomb, he was still thinking about a fission bomb, because the feasibility of the fusion bomb—the hydrogen bomb—was by no means clear, and meanwhile he thought he saw ways to do fission bombs in yield ranges immensely exceeding the scale of anything yet exploded. There were about a hundred people in the Theoretical Division, and Ted's confidence, at age twenty-four and after Berkeley, was not such that he thought his voice would ever amount to much among them. Established stars, like Stanislaw Ulam, the mathematician, and Edward Teller, the physicist, were at work on the enigma of the hydrogen bomb. Taylor's first assignment, by contrast, had been to calculate the possibilities of making a somewhat smaller version of the old bomb that had been exploded over Nagasaki. He became absorbed with fission, with its possibilities, both great and small, and he was surprised by how much he seemed to see in his own mind in comparison with how little had been done at Los Alamos since the war. When he arrived, there were no good efficiency calculations, for example. Within the Theoretical Division, there was much more interest in hydrogen bombs than there was in fission bombs. The prevailing attitude was that fission bombs were conceptually more or less a finished chapter—an assembly-line matter, no longer of great interest to the designer. Their future would be just a question of paring them, trimming them, tidying them up. Contemplating in turn each of the components of a Nagasaki-type bomb, Ted calculated the relationship of one part to another to another—the various densities, alternative materials—and he began to think how scientifically crude it was to test new or varying components all at once. Yet that had been standard procedure. The bomb was a sphere within a sphere within a sphere within a sphere. The small sphere in the center was called the initiator and was designed to give off millions of neutrons when squeezed. Around the initiator was the ball of fissile material, metallic uranium-235 or plutonium-239, in which the neutrons from the initiator would make fissions. Around the uranium or plutonium was the reflector (also called the tamper). It was made of natural uranium or some other heavy metal to prevent neutrons from getting out and to contain the explosion just long enough to prolong the fission chain reaction and produce a greater yield. Around the tamper was ordinary high explosive, the bulk of the bomb. Basically TNT, its purpose was to squeeze the uranium or plutonium from a subcritical density to a supercritical density, squeezing the initiator at the same time and creating an instant fireball. The high explosive had to be set off with something like absolute symmetry all around the sphere, or the squeeze, the implosion, would not be adequate. Two-thirds of the force of the high explosive went outward, and was lost anyway, so the implosive one-third had to be all the more nearly perfect. Mathematics had shown that charges shaped as lenses were best at starting such a process, so lenses formed the outer part of the sphere. The lenses looked like breasts, and each—there were dozens of them—had a kind of nipple to which a wire was attached. The wires ran to a common source of electricity, and this detonated the bomb. Timing was crucial. Differences of as little as a millionth of a second in the time at which lenses were detonated could affect the symmetry of the implosion, bringing it in too early on one side and thus failing to compress adequately the metal within. Timing, absolute densities of material, deviations from perfect symmetry—Ted explored each aspect of the art and decided that multiple small-yield nuclear explosions, each testing a separate aspect, would bring the level of conceptual design, as he put it, "closer and closer to the middle of things."

The idea was, if nothing else, impractical—a whole series of nuclear explosions just for one bomb. Shyly, Ted talked it over with George Gamow—an expansive, garrulous, imagistic physicist, to whose Russian warmth Ted felt drawn, as did everyone else at Los Alamos. Gamow, a progenitor of the Big Bang theory of the creation of the universe, had also postulated the mechanism for alpha-particle decay, and was the author of *Mr. Tompkins in Wonderland* and *Mr. Tomp-*

kins Explores the Atom—popular elucidations of such subjects as relativity and quantum theory, illustrated with his own drawings. Gamow had defected from Russia in the thirties. With growing interest, he listened to Ted's idea, and he reviewed calculations Ted had made in contemplation of new sizes, new shapes, new yields—all within a whole new method of approach. Gamow spoke with Teller, and with Enrico Fermi. Eventually, a meeting was called of the senior members of the Theoretical Division and its consultants, and Ted was asked to explain his concept. Norris Bradbury, the director of Los Alamos, was there, and so were Teller, Fermi, Gamow, Ulam, Konopinski—people whose names Ted had known for years, people he had wondered if he would ever meet. The situation was something like the oral examination, with the difference that Ted felt no apprehension whatever, although he was intensely excited. He was, above all, interested. He knew exactly what he wanted to say. He began by explaining that he was there to talk about experimental testing of individual phenomena inside implosion systems. He said it seemed clear that a whole task force couldn't go out to the Pacific just to test a part, so a test site on the North American Continent would be necessary. He outlined the sorts of experiments he had in mind. The meeting reacted with enthusiasm, and, of course, was followed by much critical review. The result was the first series of nuclear tests in the United States—and the establishment of the

A.E.C.'s test sites at Yucca Flats and Jackass Flats, Nevada. "You can imagine what this did for my ego," Ted would say many years later. "After Berkeley, I had really been down in the dumps. That meeting, and one or two after that, brought me up to a level of confidence that has been maintained ever since. I've never lost it."

After that meeting, Ted was given lots of time, open access to the computer, freedom of the imagination. Once in a while, he visited the divisions that actually built the bombs, and poked around in the shops there, but in his work he did not handle nuclear materials. He was a conceptual designer. All he needed was a hand calculator, a slide rule, pencils, some blank paper, some graph paper, and, from time to time, a computer output. Completing a new design, he would go to Carson Mark with a piece of graph paper in his hand—lines and numbers on it—and Mark would look at it and often say, in a way that was extremely rewarding to Ted, "Well, I'll be damned!" Ted had a desk, but was not disposed to sit. He could not think sitting down. He walked around a lot, up and down the corridors of the rambling wooden building where the wartime bombs had been designed. As he walked, his eyes swam with calculations, and he lost touch with his surroundings while nuclear devices gradually took form within his mind. If he snapped his fingers with both hands, he was thinking particularly well. If his body wiggled, he was even closer to fresh solutions. At home, he moved around a lot, too. Caro by now was accustomed to this. "When we first got married, that had been the most difficult thing of all—to exchange confidences with a moving target."

Ted loved to hike, not only indoors but out, and on weekends he would sometimes go to the Sangre de Cristo Mountains, above Santa Fe, across the Valley of the Rio Grande. He walked on a trail on the ridgeline, at twelve thousand feet, conceiving bombs. Any number of them fell apart in his mind, or on the scraps of paper back at his desk, but when they did not fall apart they were shown first to Mark, then to Jane Hall, a physicist who served as the link to the military, then to Duncan MacDougall, leader of the GMX Division—the division that dealt with high-explosive components. A design that got past these three was then put on the agenda of a meeting of the Fission Weapons Committee. Ted was required to do what he called "a selling job" before the committee, flogging his new bomb. When he was successful, the metal core and the high-explosive components were made separately and shipped to Eniwetok or Nevada, where they were assembled.

Caro knew only in a general way what Ted was doing, and she was, in her words, "sort of shocked by it—considerably so—although Ted had found a way to see that it was a good thing." She decided that she could do nothing about it, so she tried to put the matter out of her mind. This was somewhat difficult, because explosions occurred frequently at Los Alamos. The GMX Division, which had a huge machine that could X-ray an implosion, tested high-explosive systems in the canyons that splayed the town. Los Alamos was ringed with antic signs—"DANGER EXPLOSIVES KEEP OUT."

Down a slope from the Taylors' house was a culvert, and Ted decided that if a red alert should come, and there was no time to seek better shelter, his family would huddle in the culvert. In time, he discussed bomb attacks and civil defense with his children—radiation, fallout, shock waves, nuclear-weapons effects. This was too much for Caro. "Ted is a good talker. If he wants me to believe something is bad, I believe it. I still expect the worst when I hear a siren. We knew too much about all these things." A forest fire once threatened Los Alamos, and Ted and others went out to fight it with shovels. The wind shifted and the fire crowned over their heads. Ted had never before been so frightened. He had to run for his life.

Los Alamos was set in a beautiful place, with its high altitude, its clear air, its big pines, and the forested mountainsides rising above its mesas. Ted and Caro looked out a window one morning and saw a black bear in a tree. There was plenty to do beyond the town gates—ski, climb, fish for trout, camp beside a stream beneath the ponderosas. Views from the streets of the town reached forty miles to the snow-covered Sangre de Cristos. Obviously, more than remoteness had drawn Robert Oppenheimer to establish a wartime scientific laboratory around a group of log buildings called the Los Alamos Ranch School, which he had often visited on his vacations in New Mexico. "Shut in by a gate, shut out by a gate, your perspective was limited, though," Caro has recalled. "Los Alamos was a town inside a fence. At a dinner party, the men went off into a corner—more so than ordinarily. You didn't think about it, but there was something that was missing. Not long after we arrived, churches and privately owned stores and such were for the first time permitted to come in there. Churches suddenly sprang up of all sizes and kinds—about every denomination you can think of. It must have been an indication of something. Los Alamos was a nervous collection of middle-aged young people. The average age there was eighteen. The average adult age was thirty-five. You were children of the government, paying subsidized rents and living in prefabricated homes. There were wild drinking parties, marriages ending, and so forth. Ted did not particularly notice this. For him, it was an exciting community. Hans Bethe came to dinner, in hiking boots. Sometimes, Ted would go with Mr. Fermi on hikes in the mountains above the town. It was hard for a person to know how Ted was coming along. A person's confidence had been a little shaken, maybe. You worry just a little bit about your husband professionally, you know. If he could entertain Mr. Fermi for a whole walk, for a whole morning, I figured he must be a good listener or he must have something to say. On the evenings when we happened to go out, I was impressed by the people who seemed to want to talk to him in corners. Ted would say that a device of his was to be set off in Eniwetok, and I'd know he was doing O.K."

In the fall of 1950, Ted made a long visit with two others to Washington as a kind of emissary to the Pentagon—to brief and to be briefed. He was twenty-five years old and had been a junior-grade lieutenant in the Navy, and now suddenly he found himself in the supreme palace of the military, being ushered around by fleet admirals and four-star generals who treated him with the kind of reverential respect they might have shown a legitimate son of Thor himself. Meals were served in the Flag and General Officers' Mess. First, the Pentagon acquainted the scientists with a summary of what might happen if the United States were to attack the Soviet Union. Then an entire day was spent reviewing what would happen if Russia—whose first nuclear bomb had been exploded a year earlier—were to attack the United States. The generals making the presentation succeeded in frightening Ted. The drift of their argument was that Russia would win easily, that Russia had only a small nuclear stockpile but with conventional arms would overwhelm Western Europe, and meanwhile the Russians were safe enough at home, as a result of an intensive program of civil defense. For six weeks thereafter, Taylor looked through stereoscopic glasses at three-dimensional photographs of bits and pieces of the Soviet infrastructure—a refinery here, an assembly plant there. Pentagon target analysts drew circles with compasses around Soviet military bases, industries, cities. The pictures had been made by the Germans during the Second World War. Taylor's role was to estimate how many kilotons would be required to remove something from a picture. The Pentagon wanted to figure out what, cumulatively, was needed to destroy the Soviet Union.

One result of these discussions was a military request for an extremely high-yield fission bomb. The generals thought that a hundred kilotons was the upper limit of possibility. Ted said he felt he could do a lot better than that. Really? The generals were under the impression that a fission bomb beyond that limit would have to contain so much uranium or plutonium that it would go critical and fizzle before achieving detonation. Ted said he felt the problem could be avoided. The generals wished him luck, saying they hoped for a fission bomb with a yield high enough to enhance what they called the country's deterrent posture. Ted went back to Los Alamos and designed the Super Oralloy Bomb.

At about the same time, work on the hydrogen bomb had reached a level of frustrating bafflement, if not paralysis. The theoretical potentialities were clear. The problem was how to ignite the thermonuclear material (liquid deuterium or whatever) in a way that would cause a fusion-reactive explosion. Very high temperatures were required, and these existed in an exploding fission bomb. But just to set one off inside a barrel of deuterium would not do, because the energy of the fission explosion would dissipate too fast as radiation or be drained off by electrons, which would simply whirl faster and accomplish nothing. One day, at a meeting of people who were working on the problem of the fusion bomb, George Gamow placed a ball of cotton next to a piece of wood. He soaked the cotton with lighter fuel. He struck a match and ignited the cotton. It flashed and burned, a little fireball. The flame failed completely to ignite the wood, which looked just as it had before—unscorched, unaffected. Gamow passed it around. It was petrified wood. He said, "That is where we are just now in the development of the hydrogen bomb."

S.O.B., the Super Oralloy Bomb, in time was detonated at Eniwetok, and—in Ted's mind, anyway—S.O.B. solved permanently the problem of the high-yield bomb (of whatever type), for anything larger seemed redundant. It was exploded at high altitude. As soon as the generals saw the fireball, they knew they had got what they wanted. S.O.B. was—and it still is—by far the largest-yield pure-fission bomb ever constructed in the world. I once asked Taylor how much, if any, plutonium it had in it, and he said, "No comment." I asked him how much nuclear material was in it, and he said, "A lot."

"Can I say 'toward a hundred kilograms'?"

"I wouldn't say anything."

"What can you say about the yield?"

"It was in the megaton range."

In the parlance of weaponry, "the kiloton range" is the phrase used to describe fission bombs. "The megaton range" is the phrase used to describe hydrogen bombs. "Megaton range" has only been used once to describe a fission bomb.

A bomb test was an attractive aspect of work at Los Alamos—a business trip, sometimes to the South Pacific, the witnessing of an unforgettable spectacle—and there was a pecking order about who got to go. Ted did not see S.O.B. explode. He went to few bomb tests in the early years, because in the corporate scale of things he was so low. He was a forty-five-hundred-dollar-a-year designer with no Ph.D. He was a research assistant, not a member of the staff. Others, most of the others, came first. An exception was made for a test in Nevada. Ted had designed a bomb called Scorpion, which contained a component radically different from anything that had preceded it, and he was invited to be present at the explosion that would ratify or disprove his invention. The component was the reflector, which was ordinarily made of metals heavier than steel. The reflector traditionally added a considerable amount to the total weight of a bomb. Wandering around the corridors, musing about reflectors, Ted had begun to contemplate how light they might possibly be—how to make a single long stride toward a new generation of lightweight bombs—and his thoughts started through the periodic table from hydrogen, the lightest element, upward. Hydrogen, helium, lithium, beryllium . . . Beryllium was a compact, fairly dense collection of light atoms, lighter than oxygen, lighter than magnesium, lighter than aluminum. With regard to stray neutrons that might pass its way, beryllium had a high capacity to deflect them—a "high scattering cross-section." Being a good neutron scatterer might be worth even more than being dense. Quite a lot more. Ted went back to his papers, his slide rule, and his calculator, and began to sketch out the mathematics of Scorpion. No question, you could get some sort of explosion, but how big would it be? Some people who reviewed the concept felt that trying such a reflector would result in nothing more than a fizzle yield. Nonetheless, the reflector was fabricated—an accomplishment in itself, since beryllium is both toxic and brittle. In all other respects, Scorpion would be a familiar implosion bomb, the reflector being its only component with unproved characteristics. Made in Los Alamos, wired up in Albuquerque, it was taken to Yucca Flats, and Ted was flown there to see it. Through long, flat distances, he could see mountains in every direction. The desert floor, a huge shallow basin, was covered with sage. Elevations gently rose toward the barren foothills of the mountains. The Air Force was already making a romantic film about the place, with a narrator saying,

"This is the valley where the giant mushrooms grow, the atomic clouds, the towering angry ghosts of the fireballs." Such presentations would eventually sicken Ted when he came upon them, just as he would be sickened, on visits to Strategic Air Command bases, by signs that said, "WAR IS OUR BUSINESS." Now, though, it was all new to him, and he thought of Scorpion only as a device that would go off in a desert—an excitement, a spectacle, an investigation of physical phenomena. He spent what he remembers as "essentially no time" contemplating that Scorpion might be reproduced as a weapon for killing human beings. He marvelled instead at the exceptional clarity of the Nevada air and how the distances he could see were so great they were deceiving. Scorpion's tower, from across the desert, looked like a wire stuck into the ground. The tower was three hundred feet high. The height had to be greater than the expected radius of the fireball, so that the shock wave breaking away from the fireball would bounce off the ground and push the fireball upward, preventing it from picking up debris and causing unnecessary fallout. The top of the tower was an eight-by-eight-foot cab that had sides of corrugated iron, a gable roof made of sheets of iron, and an iron-grating floor. Ted went up the tower to look at Scorpion before it was fired—a dark object in the center of the cab. The tower itself contained vertical piping that was used to conduct specific types of radiation to instruments that would measure their intensity milliseconds before the instruments themselves would be destroyed. The explosion was delayed a day while technicians tried to get a rat out of a pipe. Ted, in the course of the wait, found a parabolic mirror with a small hole in it at the bottom of its concavity. Facing it to the sun, he determined where, behind the mirror, the collected rays came to focus. He attached some stiff wires to the mirror and shaped them so that they would hold a Pall Mall with one end at the focal point. Finally, at dawn the next day, Scorpion was detonated. Dawn was good for photography. Yield was measured by the rate at which a fireball grew. Ted was far across the desert, watching—hoping (he would always be highly optimistic about his bombs) for twenty kilotons, while some others were prophesying a much smaller yield. People who have worked for decades at Los Alamos have said that you can read all there is about tanks, ships, and buildings disappearing in vapor but the experiential fact is that you don't know what a kiloton is until you see and, in a sense, feel one. Los Alamos people were always taken aback when they first went to a test site and saw a bomb explode. The light came first, and then the waves and waves of heat. Many seconds later came the sound, which varied from a dull thud to a sharp crack. Scorpion, on its day, spread out into the sky in a way that indicated at once a yield in the range its designer wanted—a fantastic hint of how light and compact a nuclear-explosive device could ultimately be. Fifteen seconds after Scorpion flashed, Ted reached down to the parabolic mirror beside him and took from behind it a smoldering Pall Mall. He drew in a long, pleasing draught of smoke. He had lit a cigarette with an atomic bomb.

To produce enough electricity to keep Yonkers going for a year, a light-water nuclear reactor would make, as a by-product, just about enough plutonium to obliterate Yonkers. Yonkers, in population, is an average American city. Something like two hundred thousand people live there, on eighteen square miles of land. The ratio of plutonium to electric power would be similar for any community. If a risk is present, it is taken to get electricity, and the philosophical drift beneath all this seems to be that the electricity is very much worth the risk. For almost a hundred years, parts of mankind have been using electricity commercially, and in that time electricity has evolved from a novelty into a necessity, and the possibility of running short of it, which would have been a meaningless concept a century ago, is now quite properly called a crisis. So much would collapse without it. Ways of making it, alternative to nuclear reactors, are diminishing. "Do we really need this new source of energy?" Glenn Seaborg said when he was chairman of the Atomic Energy

Commission. "Not only do we need nuclear power, but this source of energy has, historically speaking, been discovered just in the nick of time." If self-preservation is elemental in human nature, so is a capacity for accepting risk when something is wanted enough. Portuguese sailors travelled fifteen thousand miles to fill small wooden boats with oranges and cinnamon. The cravings of the race are closer to home now, and the people of Yonkers are locked in solid to a requirement for electric power, perhaps at risks that would beggar the exploits of the merchant Portuguese.

Mason Willrich, a professor of law at the University of Virginia, has for many years been what might be called, in a legal context, a nuclear scholar, and in his *Global Politics of Nuclear Energy* he points out that five hundred years ago in England it was a capital offense to burn coal. When coal was burned, the gas that came off it was shown to be poisonous. It seemed clear enough even then that the air of a coal-burning London could kill human beings. So coal was put on the forbidden list. There was plenty of wood anyway. The world had once had twelve billion acres of virgin forest, and a good deal of it was left, although great deserts had been created where forests once stood in Asia. England burned wood until her hills were bald. By the nineteenth century, an energy crisis had developed, and some alternative source needed to be put to use, in the nick of time. Willrich continues:

> The transition from wood to coal awaited the

> invention of a variety of coal-burning machines and the development of a demand for energy that was urgent enough to reduce, in people's minds, the adverse side effects of coal-burning from a lethal menace to a tolerable nuisance. Coal powered the Industrial Revolution initially, put the horse out to pasture, and, depending on one's view, helped to free the slave or chain him to a machine. Energy consumption increased, standards of living rose sharply, and a belching smokestack became a symbol of progress.

Willrich was once a co-pilot in a B-47. Part of his job was to arm the nuclear bombs the plane carried. To practice what were known as "manual in-flight insertions," he had to crawl through a passageway and into the bomb bay, where he twisted a cylindrical device that worked more or less like a combination lock and placed related masses of nuclear material in a position to be fired. He imagined getting his arm caught and being pulled out of the plane and falling to earth with his arm stuck in the bomb. It was a daydream at first, but it turned into a recurrent nightmare—one that from time to time he still has. A bomb once did fall out of the plane, but the plane happened to be sitting on the ground. The bomb was a huge, clumsy, early-generation thermonuclear bomb, almost three hundred cubic feet stuffed into a steel jacket, and it hit the runway with a great thud and made a deep dent in the tarmac. Willrich has been worried about nuclear power ever since his days in the Strategic Air Command. Before beginning to teach law, he worked some years in the Arms Control and Disarmament Agency. He is particularly worried that bombs might be made in a clandestine operation by a nation, a group, or an individual. He does not believe that the problem is insoluble, that nuclear material is uncontrollable. He is, in fact, a proponent of civilian nuclear power.

> The invention and perfection of the internal combustion engine extended the Industrial Revolution to its logical conclusion. Oil rapidly challenged, then supplanted, coal as the source of primary energy for transportation—with a revolutionary impact on man's mobility. But even as the Industrial Revolution entered into full sway in the technologically advanced nations, the energy resources and technology were being developed on which "post-industrial" societies could rest.
>
> Electricity was introduced at the end of the nineteenth century. The transition from an industrialized to a post-industrial society can be viewed as partial cause and effect of the electrification of more and more areas of human activity. For this reason, the process of transition will be arrested unless electric power remains cheap and becomes more abundant in the future. Thus, hope rests ultimately on nuclearization of electric-power production.

A nuclear power plant is a contemporary demonstration of how far people not only can but will go to get what they want or need. This, for example, is Indian

Point No. 2—a Consolidated Edison reactor beside the Hudson River. The containment structure looks vaguely like the Jefferson Memorial—a simple, stunning cylinder under a hemispherical dome, all in white reinforced concrete. Its diameter is a hundred and thirty-five feet. The dome and the side walls are from three and a half to four and a half feet thick and are lined with steel. The welds in the steel are bracketed with channels in which positive pressure must be maintained, so that nothing can pass through a crack. A human being, to enter, has to pass through an air lock and sign in at a security desk. The feet on the desk are those of the security man. He is quite young. His hair falls below his shoulders. He wears white cotton shoe covers and white coveralls. He looks up slowly from his book. He is reading *Portnoy's Complaint.* The reactor reaches only about a third of the way up into the interior space, which must be voluminous in order to contain immense outpourings of steam should an accident happen. The building is two hundred and fifteen feet tall. The main interior deck is a hundred and sixty-five feet below the center of the dome. The first moment of looking upward is more impressive than the analogous moment in the Guggenheim Museum. Spiralling within the dome is a sprinkler system that could launch a ship. Around the reactor is a pattern of shielding walls within walls, each about as thick as the outer containment wall itself—blockades of concrete set up to stop particles so small they would be invisible at a billion magnifications. Concentric with the outer wall is a cylindrical shielding wall that supports a polar crane, which handles the reactor parts—lifts the lid (the Reactor Vessel Head), lifts the shafts that guide the control rods (the Upper Internals Package), moves machinery around the room. Completely within the polar-crane wall is a neutron-shield wall, six feet thick. The reactor and these two protective walls are set in the Reactor Cavity Pool, an irregular rectangle that is also, in places, six feet thick. It can be flooded to a depth of thirty-five feet. Wherever any wall is penetrated by pipes, tubes, or wires, it is backed by an additional concrete barricade. The implication of this labyrinth is that something extremely solemn exists within, and one strains for the opportunity to stand on the neutron-shield wall and look down into the reactor core. Fortunately, that is possible at the time of this visit, for Indian Point No. 2 has not yet gone critical for the first time. Fuel assemblies of the type used here have burst their cladding in another reactor. The fuel assemblies here must be removed and sent back to the fabrication plant for revision. So Indian Point No. 2 is all built and ready to go, but the reactor's Upper Internals Package has been set to one side and the core is exposed to view as it never will be again. The reactor vessel, also cylindrical, is only fifteen feet in diameter. It is about three times as tall. It nestles in the center of the building like an egg in a large covered basket. When the reactor's top comes off during actual operation and used fuel elements are removed, the core, under deep water, will shimmer with Cherenkov effect. The sides of the reactor are of steel just under six inches thick. The fuel rods, now visible within, are twelve feet high and less than half an inch in diameter —tall stems, standing parallel to one another in their assemblies to form a forest clad in zirconium alloy. Thirty-nine thousand three hundred and seventy-two fuel rods stand there together. Among them are fifty-three rods made of a silver-indium-cadmium alloy coated in stainless steel. These—the control rods, the "poison" rods—in their combined elements have a capacity to absorb neutrons and stop a fission chain reaction. They can also start one, for when the reactor is operating they are to be surrounded in the core by so much uranium that a fission chain reaction would develop uncontrollably if they were not there. When the poison rods are drawn slowly upward, by drive mechanisms in the Reactor Vessel Head, the uranium in the fuel rods will go past the point of criticality, neutrons by the octillion will start jumping around in the forest, and the temperature of the core will rise. Water flowing among the fuel rods will become extremely hot. The poison rods will move out or in just enough to hold the water temperature at six hundred degrees Fahrenheit. The reactor vessel holds water under pressure and is thus an enormous pressure cooker. Flanges join the upper and lower parts and are held together by dozens of steel studs. The studs are seven inches in diameter and five feet long. While the lid is being closed, the studs are stretched, so that when they are released the seal will be that much tighter. Not tight enough, though—so two great hoops, made of silver-plated Inconel, go into grooves in the interface of the flanges. Nothing should get past these O rings, as they are called. They cost fifteen thousand dollars the pair. The entire reactor unit goes for two hundred and fifteen million. Westinghouse builds it and hands you the key.

Several hundred people work at Indian Point. A sign on the inside of the containment wall says, "CALL HEALTH PHYSICS IF ALARM ANNUNCIATES." Another sign says, "NO EATING, NO DRINKING, NO SMOKING." Someone has added, "NO BREATHING, NO MASTURBATING." On the neutron-shield wall someone has lost a game of ticktacktoe. John Makepeace, the plant engineer, proud of his construction, shows it off from bottom to top. He leads the way down a steel ladder and into a cramped space directly underneath the reactor vessel, where hundreds of in-core flux detectors penetrate the reactor, and where the building as a whole rests on limestone, dolomite, and the People's Republic of China. From the roof of the plant, the view is a bold one of the promontories of the Hudson Highlands. The sun is bright. The river is blue. The air is clear. The wind is thrashing an American flag. Six cables lead away from the roof—up to a tower and out to the world. Half the power from Indian Point No. 2 will go to the Consolidated Edison substation at Sprain Brook,

Yonkers. It takes two hundred kilowatt-hours to make four aluminum lawn chairs. It takes two kilowatt-hours to make six aluminum beer cans. A psychiatrist must be air-conditioned or his listening power is not negotiable. It is twenty-odd years since the atom was declared peaceful. Fifteen or twenty more years and the A.E.C.'s big driver reactors at Savannah River, South Carolina, and Hanford, Washington, the sources of plutonium for bombs, will be surpassed in cumulative plutonium production by the reactors of utility companies, making plutonium for private stockpiles.

I ONCE asked Ted Taylor if he was at all worried about people making hydrogen bombs in their basements and, if so, how they might go about it.

He said, "I can't tell you anything at all about that except that my opinion is that a homemade H-bomb is essentially an impossibility. One can't even hint at the principles involved, beyond saying that it requires heating some material up to a terribly high temperature, which is why it is called a thermonuclear bomb. There are by now several thousand people who know how this is done, so the secret of the H-bomb will out somewhere along the line, but, even when it does, the fact remains: to make an H-bomb is not a basement operation. The project would take a large, well-organized group of people a great deal of time. The secret, incidentally, is not a matter of materials. It is a matter of design."

The design was hit upon by Stanislaw Ulam and Edward Teller in 1951. In the pages of a patent application they were described as the bomb's "inventors." After a long period of getting nowhere—an effort by many scientists, under considerable pressure from Washington—Ulam one day asked Teller to sit down in private with him and listen to an idea. They closed the door of Teller's office at Los Alamos and talked. Teller was much impressed with Ulam's idea and at once thought of a better way to do the kind of thing Ulam had in mind. The two men came out of the room with the answer to the problem of the hydrogen bomb. The rest was detail, albeit on a major scale—computer calculations, design, fabrication. The better part of two years went by before a task force was ready to go—in the fall of 1952—to Eniwetok to test the theory.

Not all Los Alamos theories could be tested. Long popular within the Theoretical Division was, for example, a theory that the people of Hungary are Martians. The reasoning went like this: The Martians left their own planet several aeons ago and came to Earth; they landed in what is now Hungary; the tribes of Europe were so primitive and barbarian that it was necessary for the Martians to conceal their evolutionary difference or be hacked to pieces. Through the years, the concealment had on the whole been successful, but the Martians had three characteristics too strong to hide: their wanderlust, which found its outlet in the Hungarian gypsy; their language (Hungarian is not related to any of the languages spoken in surrounding countries); and their unearthly intelligence. One had only to look around to see the evidence: Teller, Wigner, Szilard, von Neumann—Hungarians all. Wigner had designed the first plutonium-production reactors. Szilard had been among the first to suggest that fission could be used to make a bomb. Von Neumann had developed the digital computer. Teller—moody, tireless, and given to fits of laughter, bursts of anger—worked long hours and was impatient with what he felt to be the excessively slow advancement of Project Panda, as the hydrogen-bomb development was known. Kindly to juniors, he had done much to encourage Ted Taylor in his work. His impatience with his peers, however, eventually caused him to leave Los Alamos and establish a rival laboratory at Livermore, in California. Teller had a thick Martian accent. He also had a sense of humor that could penetrate bone. Dark-haired, heavy-browed, he limped pronouncedly. In Europe, one of his feet had been mangled by a streetcar.

Ulam was a Pole and had no inclination to feel thunderstruck in the presence of Hungarians, whatever their origins. His wife was vivacious and French. He worked short hours. He was heroically lazy. He was considered lazy by all of his colleagues, and he did not disagree. The pressures of the Cold War were almost as intense at Los Alamos as the pressures of the war that preceded it, but these pressures were resisted by Ulam. Mornings, he never appeared for work before ten, and in the afternoons he was gone at four. When Enrico Fermi organized hikes on Sundays, Ulam went along to the foot of the trail. Fermi, Hans Bethe, George Bell, Ted Taylor—up the talus slopes they went while Ulam sat below and watched them through binoculars. Many years after the first thermonuclear bomb had been successfully tested, Ulam's secretary cut out and tacked to a bulletin board a cartoon in which two cavemen were talking about a third caveman, who was standing off by himself. The caption was "He's been unbearable since he invented fire."

The object that had been sent out to Eniwetok was distinguished by its plainness and its size. It was a cylinder with somewhat convex ends. It was twenty-two feet long and five and a half feet in diameter. It was the result of Project Panda, and it was called Mike. It looked something like the tank on a railway tank car, and it weighed twenty-one tons. Inside it was at least one fission bomb, and a great deal of heavy hydrogen. Mike was placed in a building with metal siding which had been constructed for the purpose on an island called Elugelab, in the northern sector of the atoll. After Mike had exploded, nothing whatever remained where the island had been but seawater. The island had disappeared from the earth. The yield of the Hiroshima bomb had been thirteen kilotons. The theoretical expectation for Mike was a few thousand kilotons—a few megatons. The fireball spread so far and fast that it terrified observers who had seen many tests before. The explosion, in the words of Ted Taylor, who was not there, "was so huge, so brutal—as if things had gone too far. When the heat reached the observers, it stayed and stayed and stayed, not for seconds but for minutes." The yield of the bomb was ten megatons. It so unnerved Norris Bradbury, the Los Alamos director, that for a brief time he wondered if the people at Eniwetok should somehow try to conceal from their colleagues back in New Mexico the magnitude of what had happened. Few hydrogen bombs subsequently exploded by the United States have been allowed to approach that one in yield. The Russians, however, in their own pursuit of grandeur, eventually detonated one that reached just under sixty megatons—more than four thousand times the explosive yield of the Hiroshima bomb. Taylor guessed that if the Russians had wrapped a uranium blanket around it they could have got a hundred megatons, but he imagined they were afraid to go that far. Whatever the size of the big bombs—ten, sixty, or a hundred megatons—they had begun to dismay him long before they were tested. (In seven years at Los Alamos, he would work on the design of only one hydrogen bomb.) He was even sorry that he had designed the Super Oralloy Bomb, for the belief he had once held in the "deterrent posture" of such huge explosions had eventually dissolved in the thought that if they ever were used they would be "too all-killing"—that the destruction they would effect across hundreds of square miles would be so indiscriminate that the existence of such a weapon could not be justified on any moral ground. He reached the conclusion that an acceptable deterrent posture could only be achieved by making small bombs with a capacity for eradicating specific small targets. The laboratory's almost total emphasis in the other direction—toward the H-bomb—bothered him deeply. He began wondering just how small and light a nuclear explosive could be—how much yield could be got out of something with the over-all size of a softball. With George Gamow, he wrote a scientific paper called "What the World Needs Is a Good Two-Kiloton Bomb."

In 1953, Taylor was sent by Los Alamos at full pay to Cornell, to spend a year and a half getting his Ph.D. His mentor there was Bethe, who had long since become a close friend and counsellor—a relationship that continues. When Taylor returned to Los Alamos, he resumed the conception of a number of bombs whose names unmistakably indicate the direction of his effort: Bee, Hornet, Viper, the Puny Plutonium Bomb. The test of that last one was called "the P.P. shot," and it was the first known complete failure in the history of nuclear testing. "Now you're making progress," Fermi said. "You've finally fired a dud." The bomb had plenty of high explosive around an amount of plutonium so small that it remains secret, for the figure is somewhere near the answer to the root question: How small can a nuclear bomb be? In the absence of precise numbers, the answer would have to be: Pretty small. Fiddling around on this lightweight frontier, Taylor once designed an implosion bomb that weighed twenty pounds, but it was never tested.

Studying ordinary artillery shells, he replicated their external dimensions in conceiving fission bombs that could be fired out of guns. Being longitudinal in shape, these were not implosion bombs, of the Nagasaki type, but gun-type bombs—the kind that had been dropped over Hiroshima. The basic idea was to fire one piece of metallic uranium down a shaft and into another piece of metallic uranium, turning what had been two subcritical masses into one supercritical mass that would explode. Taylor called this exercise "whacking away at Hiroshima," and he performed it successfully, but he was not much interested in gun-type bombs. The Hiroshima bomb, which had been designed by a committee, was overloaded with uranium, and Taylor's summary description of it was that it was "a stupid bomb." Possibilities were so much greater in implosion systems. The Nagasaki bomb's nuclear core had been designed by Robert Christy, who taught physics and astrophysics at the California Institute of Technology, and returned frequently to Los Alamos as a consultant. Taylor, in his own words, would "light up" when he found that Christy had come to town. Christy showed great interest in what Taylor was trying to do, and gave him much encouragement.

As the bombs grew smaller, the yield did, too. So the obvious next step was to try to get the yield back up while retaining the diminutions of size and weight. Taylor incorporated an essentially new feature that might do just that. It was put into Bee and Hornet. At the sound and the sight of each of them—the big fireball, the loud bang—he knew at once that the feature had worked.

"What feature?" I asked him once.

"I can't say," he answered. "So far as that part of the discussion goes, we have come to a dead end."

Freeman Dyson, one of the preeminent theoretical physicists in the world, has not worked at Los Alamos but has had occasion in his career to review closely the work Ted Taylor did there. "His trade, basically, was the miniaturizing of weapons," Dyson has said. "He was the first man in the world to understand what you can do with three or four kilograms of plutonium, that making bombs is an easy thing to do, that you can, so to speak, design them freehand." Taylor's colleagues at the laboratory came to regard him as being "halfway between an inventor and a scientist." This is how Marshall Rosenbluth remembers him. Rosenbluth, who, like Dyson, is now at the Institute for Advanced Study, in Princeton, was at Los Alamos through the same years Taylor was. Rosenbluth worked principally on the thermonuclear bomb. "Ted was not a typical physicist working out little mathematical problems," Rosenbluth has said. "He thought more qualitatively. He made many inventions, and did a pulling together of the physics necessary for them. He did not spend his time working on the most esoteric of physics points." Once, on a visit to Los Alamos, I asked Carson Mark, director of the Theoretical Division, if he would tell me what Taylor's particular thumbprint had been, as a designer and as a physicist. Mark said, "There was a need for different kinds of physics. There was a need for different kinds of contributions. Many did long computations that took weeks. Ted did not. New ideas don't come this way. There were problems in physics which took months and months to solve, resulting in benchmark papers in the *Physical Review,* such as a study of neutron scattering from some particular material. Such things require a great deal of careful work. You need it. You value people who can do it right. Ted did not do much of this. Ted's papers were shorter, more qualitative—physics sketched out but not extensively explored. Figures were needed. Ted would guess. For exact figures, a man-year's work might be involved. Ted would not be doing that work. His style was a flair for qualitative sketching of a complicated process. He was conceptual. His numbers were reasonable but were not exact. With intensity, he thought outside the prescribed context. Others could answer questions if you asked them, but they did not keep thinking of so many unlikely things. Marshall Rosenbluth and Conrad Longmire, for example, were strong physicists of a breadth and depth greater than Ted's. Ted's curiosity, his prying, his imagination—a combination equally valuable—exceeded theirs." Mark looked pensive for a while, and found an afterthought. "Ted may even have not been the most imaginative," he said. "We've had some real nuts around here."

While Ted's bombs grew smaller, some of his other ideas grew to epic proportions. He spent a lot of time walking aimlessly from corridor to corridor thinking about the slow production of plutonium. The A.E.C.'s plants at Hanford and Savannah River were literally dripping it out, and Ted thought he saw a way to make a truly enormous amount of plutonium in a short time. He wanted to wrap up an H-bomb in a thick coat of uranium and place it deep in arctic ice. When it was detonated, the explosion would make plutonium-239 by capturing neutrons in uranium-238—exactly what happened in a reactor. The explosion would also turn a considerable amount of ice into a reservoir of water, which could easily be pumped out to a chemical plant on the surface, where the plutonium would be separated out. Why not? Why not make tritium in the same way? Tritium, the heaviest isotope of hydrogen (one proton, two neutrons), is the best fuel for a thermonuclear explosion, and the most expensive (eight hundred thousand dollars a kilogram). Tritium is everywhere—in the seven seas, in the human body—but in such small proportions to ordinary hydrogen that collecting tritium in quantity from the natural world is completely impractical. So it is made, slowly, in production reactors. Ted wanted to do it a short way. Put a considerable amount of lithium around a thermonuclear bomb and emplace it under ten thousand feet of ice. Boom. An underground lake full of heavy isotopes. "That idea did not fly," said Carson Mark, in summary. "It properly received a lot of exploratory thought. It was a good idea. It would work, but it was too hard to do." These arctic ideas of Ted's became known as MICE—megaton ice-contained explosions. He found a serious supporter in John von Neumann, who was by then an A.E.C. commissioner. Von Neumann died two years later, in 1957, and the support died with him. Alternatively, Ted wanted to spread out on the ground somewhere a uranium blanket four hundred feet square. Then he would detonate a thermonuclear bomb in the air above it. Instantly on the ground there would be tons of plutonium. That idea did not even crawl.

Ted's imagination was given limited assignments as well as unlimited freedom. He was, after all, working for the government, and, as Marshall Rosenbluth remembers those days, "admirals and generals were forever calling up begging for appointments." One time, for example, Ted was asked to see how well he could do "in a certain yield range" in terms of "high efficiency, high compressions, high criticality"—no fancy innovations, just the best implosion bomb he could make within the parameters given. The result was Hamlet, the most efficient pure-fission bomb ever exploded in the kiloton range.

Driving around Los Alamos with him once, when I went along on a visit he made there in 1972, I asked him what he had done to occupy himself during the flat periods between projects, the lulls that would come in any pattern of conceptual work. He said, "Between bombs, we messed around, in one way or another. We bowled snowballs the size of volleyballs down the E Building corridor to see what would happen. We played shuffleboard with icicles." He supposed it helped relieve the tension, of which there was a fair amount from time to time. During the strain of preparation for the Mike shot, for example, a well-known theoretical physicist picked up an inkwell, threw it at a colleague, and hit him in the chest. He could be excused. His job was to make sure that the hydrogen bomb did not ignite the atmosphere. After the Livermore laboratory began making bombs in competition with Los Alamos, a rivalry developed that was at least as intense as the football rivalry between, say, Michigan and Michigan State. Each laboratory had its stars. Johnny Foster was the fission-bomb star of Livermore. Groups of scientists from the one laboratory would attend the other's bomb tests, and there was a distant sense of locomotive cheering in the air, of chrysanthemums and hidden flasks. If a Livermore bomb succeeded only in knocking off the top of its own tower, or a Los Alamos bomb was a dud, no one actually cheered, but some people felt better. Once, at Eniwetok, somebody decided to steal Livermore's flag, which was pinned to a wall in Livermore's barracks and included in its heraldry a California golden bear. From the central flagpole, Headquarters, Joint Task Force Seven, Eniwetok Atoll, the flag of Rear Admiral B. Hall Hanlon, commander of the task force, was removed in the dead of night. Hoisted in its place was the Livermore bear. The Admiral's flag was then pinned to the wall in the Livermore barracks. In the morning, Admiral Hanlon reacted as expected, personally yielding four kilotons, one from each nostril and one from each ear. What is that bear doing on my flagpole? Where is my flag? Where? God damn it, where? Captains, colonels were running around like rabbits under hawk shadow —and, of course, they found the Admiral's flag. Los Alamos had triumphed without a shot being fired.

An explosion, however large, was a "shot." The word "bomb" was almost never used. A bomb was a "device" or a "gadget." Language could hide what the sky could not. The Los Alamos Scientific Laboratory was "the Ranch." Often, it was simply called "the Hill." An implosion bomb was made with "ploot." A hundred-millionth of a second was a "shake"—a shake of a lamb's tail. A "jerk" was ten quadrillion ergs—a unit of energy equivalent to a quarter of a ton of high explosive. A "kilojerk" was a quarter of a kiloton. A "megajerk" was a quarter of a megaton. A cross-section for neutron capture was expressed in terms of the extremely small area a neutron had to hit in order to enter a nucleus—say, one septillionth of a square centimetre—and this was known as a "barn." Two new elements—numbers 99 and 100—were discovered in the debris resulting from Project Panda, the Mike shot. Some wanted to call element 99 pandamonium. The name it got was einsteinium.

Conversations were more likely to be in an idiom of numbers, though. Numbers, volumes, densities were the stuff of working thought, and of daydreams as well. Ulam announced one day that the entire population of Los Alamos could be crammed into the town water tower. Taylor figured out that the Valle Grande, a huge caldera in the mountains above Los Alamos, had been created by a thousand megatons of volcanic explosion. His conversation to this day is laced with phrases such as "of the order of" and "by a factor of," and around the top of his mind runs a frieze of bizarre numbers. He will say out of nowhere that his wife has baked a hundred and eighteen birthday cakes in the past twenty-five years, or that the mean free path of a neutron through a human being is eight inches. "The mean free path of a neutrino is greater than the diameter of the earth. They go right through the world." He says there are some numbers that are so large or so small that they are never seen, because they refer to nothing. "You never see a number larger than ten to the hundred and twenty-fourth, for example."

"Why not?"

"Because there is nothing bigger than that. That is the volume of the known universe in cubic fermis. A fermi is the smallest dimension that makes any sense to talk about—ten to the minus thirteen centimetres. That's about the diameter of an electron. Nothing we know of is smaller than that."

When I asked him how many atoms there were in his own body, he said, right back, "Eight times ten to the twenty-sixth."

We had lunch in the Los Alamos cafeteria one day with, among others, Ulam, who was now teaching mathematics at the University of Colorado but kept a house in Santa Fe and worked as a consultant at Los Alamos several months each year. Ulam began wondering aloud about the surface of a billiard ball and what it would look like if the billiard ball were magnified until its diameter were equal to the earth's. Would the irregularities of the surface be as high as the Himalayas? He decided they would. He asked Ted to come home for dinner with him and his wife, Françoise, and to bring me along. He drew a map of his neighborhood in Santa Fe and said he could not wait to lead the way because he had to leave the laboratory at four.

Ulam's house, behind a high wrought-iron gate in a warren of adobe, might have been the retreat of a minor grandee in the old quarter of Seville. It was on several levels, the lowest of which was the living room —down a few steps and into an outreaching white space that was at once expansive, under a fourteen-foot ceiling, and compact, with a tear-shaped white fireplace built into one corner. Logs were burning. They had been stood on end, and were leaning against the back of the fireplace. Cottonwood smoke was in the air. Stretched out on a large daybed during much of the evening—looking into the fire, or, with quick glances of interest or amusement, into the eyes of his wife and his visitors—was Ulam, inventor of the hydrogen bomb. A great variety of books in French and English lined the room. A grand piano stood in one corner, and on a tripod near it was a white telescope about five feet long. Ulam, always interested in the stars, had been connected with Los Alamos since 1943, and one of the earliest potentialities that occurred to him when he began work on the Manhattan Project was that nuclear-explosive force could be used to drive vehicles from Earth into distant parts of space—an external-combustion engine, fuelled with bombs. Trim, tan across his bald head, obviously well rested, Ulam was sixty-two at the time. He looked no older than Taylor, who was forty-eight. He asked about the independent research Ted had been doing for many years in the field of nuclear-materials safeguards and was much absorbed by a story Ted told him about the attempted blackmailing in 1970 of a city in Florida. The blackmailer promised not to bomb the city out of existence in return for a million dollars and safe custody out of the United States. A day later, the threat was repeated, and with it came a diagram of a hydrogen bomb. Taylor described the diagram to Ulam—a cylinder filled with lithium hydride wrapped in cobalt, an implosion system at one end of it—and nothing in Ulam's face or Taylor's manner indicated that such a diagram might not be credible. The threat, though, had been a hoax, perpetrated by a fourteen-year-old boy. The police chose not to reveal to the public that the bomb in the threat was nuclear. A judge, after sentencing the boy, suspended the sentence and put him under the guidance of two scientists in the area, saying that talent such as the boy had should be channelled in a positive direction, and not a negative one, as might happen in a prison.

Taylor asked Ulam what was new in mathematics, and Ulam said that the properties of infinity were of much philosophical interest, that there was a lot of work being done on combinatorial mathematics as it applies to biology, and that it was now possible to prove that there are some theorems that can be neither proved nor disproved. Ulam's mind wandered on to Shakespeare, to Gaudí, to Joseph Conrad—who, like Ulam, was a Pole, and first learned English when he was about twenty years old. Ulam wondered if it was possible to discern Conrad's origins in an unlabelled quantity of Conrad's prose. "I never actually read sentences," he said. "I have a good memory. I look at a page and see what is there. But I think I miss a lot." He recalled his first arrival, many years ago, at Cambridge University, and his first visit to Trinity Great Court and the college room of Isaac Newton. He said, "I almost fainted."

Before we left, Ulam found a moment to say, out of

Taylor's earshot, "I have known hundreds of people in science, and he is one of the very few most impressive and inventive. I as a boy was always reading Jules Verne. It was where I got my ideas of Americans. When I met Ted, he fitted the ideas I formed as a boy of Americans, as represented by Jules Verne. The trait I noticed immediately was inventiveness. Scientists are of different types. Some follow rules and techniques that exist. Some have imagination, larger perspectives. Often, Ted had the attitude of 'Ours not to reason why.' He was intense, high-strung, introspective. 'If something is possible, let's do it' was Ted's attitude. He did things without seeing all the consequences. So much of science is like that."

Driving away from the lights of Santa Fe and up into the mountains toward Los Alamos, Taylor fell into a ruminative mood, and eventually said, "The theorist's world is a world of the best people and the worst of possible results." He said he now saw all his work on light weapons as nothing but an implementation of "pseudo-rational military purposes." He said his belief in deterrent postures had eroded to zero. "I thought I was doing my part for my country. I thought I was contributing to a permanent state of peace. I no longer feel that way. I wish I hadn't done it. The whole thing was wrong. Rationalize how you will, the bombs were designed to kill many, many people. I sometimes can't blame people if they wish all scientists were lined up and shot. If it were possible to wave a wand and make fission impossible—fission of any kind—I would quickly wave the wand. I have a total conviction—now—that nuclear weapons should not be used under any circumstances. At any time. Anywhere. Period. If I were king. If the Russians bombed New York. I would not bomb Moscow."

To be immensely destructive, a nuclear bomb does not have to be Hamlet. It can be many times less efficient—as the Hiroshima bomb was. The making of a nuclear bomb does not require the skill and invention that went into Bee, Hornet, and Scorpion. A crude fabrication producing a small yield, or even a fizzle yield, could kill tens of thousands of people and bring tall buildings to the ground.

Ted Taylor left Los Alamos in 1956. He now lives in Maryland, in suburban Washington, not far from the Atomic Energy Commission, and in western New York, on ninety acres of forested land in the Allegheny Mountains. He is an independent technological researcher in a small company he founded, and he has supported his work and his family on grants from foundations and on contracts with agencies of the federal government, including the A.E.C. By the hundreds of thousands of words, he and his colleagues produce computer-assisted studies dealing with matters as diverse as greenhouse agriculture, pollution-control economics, and the efficiency of the United States Postal Service. As he approaches the age of fifty, he would like to forget forever the craft of nuclear weaponry, but events make that impossible for him. With the ongoing rise of civilian nuclear power comes plutonium recycle, the fast breeder reactor, a world flow of weapons-grade material in the millions of kilograms. Over his shoulder, the horizon is stuffed with thunderheads. He, of all people, knows what might be done, and how easily, with stolen uranium-235 or plutonium-239. What worries him most is that "national full-scale violence may not apply as an inhibitory force," for the so-called posture of deterrence, nation versus nation, would have no influence at all on a small group or an individual fabricating in secret a nuclear bomb. For a decade or so, he has attempted in any way he could to express his worry, in the light of his special knowledge, to the United States Congress, the Atomic Energy Commission, and the International Atomic Energy Agency, among others. He is only one of many who are equally worried, and he does not fail completely to get sympathetic attention. On the whole, though, he has been turned politely away. It is said of him that he makes the mistake of supposing that other people are as talented as he is. ("He seems to think that anybody could do it, but that is not so. If you wanted to make a bomb, you would need a Ted Taylor.") Moreover, his cautions can be an irritant to people who see themselves as the Horatiuses of the energy crisis—to an Atomic Energy Commission that, on the whole, regards itself as part of "the nuclear business," to a Consolidated Edison busy making "Clean Energy" for a cleaner environment. From some quarters, a chorus of disavowal places nuclear-materials safeguards below the threshold of reasonable worry.

"No one could just steal material and make a nuclear bomb."

"It is not possible."

"You would need your own Manhattan Project."

Ted Taylor would like to see Los Alamos or Livermore build and detonate a crude, coarse, unclassified nuclear bomb—unclassified in that nothing done in the bomb's fabrication would draw on knowledge that is secret. Certain questions have arisen, though, over what is and is not secret. Once, at the Atomic Energy Commission, Taylor was suggesting the factors of density that could be reached in metallic fissile material with certain levels of implosive force, and he was warned never to repeat what he had said in public print or in a public speech, because everything he had been saying was classified. Taylor replied that everything he had been saying he happened to have read the day before in the Encyclopedia Americana. A bombmaker could probably get along without the Encyclopedia Americana anyway. So many books contain information of similar value. Taylor's instructions to Livermore or Los Alamos would be "Lay off any sophistication altogether. Try to see what is the simpleminded way to make something that could knock over the World Trade Center. Try to see how sloppy you can get. Then set the thing off underground. Measure the yield. Put a stop to speculation about this subject." The Atomic Energy Commission has never come near endorsing such a plan.

"That piece has been overplayed."

"You would need your own Manhattan Project."

As one might imagine, Taylor has many times made simple, unclassified bombs in his mind. He has made them with uranium and with plutonium and in varying forms and styles. He has satisfied himself to the point of certainty that a homemade nuclear bomb is not an impossibility, that such an undertaking need not even be particularly difficult, and that the people who could do it are countable in an expanding number that is already in the many tens of thousands. For some years, he attempted from inside the nuclear world to influence the development of materials safeguards that would be good enough to allay his fear. He felt his efforts were unsuccessful. So, near the end of the nineteen-sixties, he decided that his views of the safeguards problem properly belonged to the public and should be vented in a public way. At universities, he began to give lectures and papers on safeguards. He contributed chapters to books with titles like *Preventing Nuclear Theft: Guidelines for Industry and Government.* He has talked to journalists, sometimes at their request and sometimes at his own. He has outlined the gist of his worry on television. The usual result of these forays is a short, local burst of public interest, with a half-life of about two days. People don't seem to care whether they blow up on Columbus Day or soon after the first of the year. The A.E.C., for its part, steadyingly assures inquirers that the problem is, first of all, under control and, second, not serious. "You would need your own Manhattan Project."

In 1972, Taylor was given a grant by the Ford Foundation to do a thoroughgoing study of the safeguarding of special nuclear materials, for publication in book form—*Nuclear Theft: Risks and Safeguards* (Ballinger)—in 1974. The study was done in collaboration with Mason Willrich, of the University of Virginia, who handled the legal and sociological aspects and what he called "the risk potential." Taylor looked after the physics, the physical whereabouts of the fissile material, the nuclear-power fuel cycle. He travelled all over the country, from M.I.T. to San Diego, West Valley to Los Alamos. At my request, he took me along. He said (and I am putting this together from fragments spoken over a number of months), "I'm an advocate of nuclear power, but there's a certain aspect of it that has to be fixed. Is it better to discuss this in the open now or later, when so much material will be around that the idea of the clandestine bomb will obviously occur to someone? The A.E.C. thinks the devices we made at Los Alamos are too complicated for clandestine manufacture by amateurs. True, our bombs were complicated. But there are much easier ways to make bombs. Getting this out into full view may accelerate action that will make the probability of a clandestine bomb less likely. Risks are associated with the use of any type of energy. The question is: Are the risks, in the light of the benefits, reasonable to take on? It seems to me wrong, and impractical, for the public at large not to be presented with the class of risks as well as with the benefits. I see no way of doing this without going into the risks. It seems necessary to be quite specific. You have to make the risks credible or people will find a way not to believe you. You can't just say, 'Bombs can be made on a scale that does not approach the scale of the Manhattan Project,' and let it go at that. Historically, public pressure is the only kind that the nuclear community responds to. The special-nuclear-materials problem is getting bigger now, and before long will increase by a huge factor. Someone will get the idea. In England, about five years ago, somebody—it was a hoax—*advertised* U-235 for sale. He got plenty of takers. For the making of a bomb, more than enough information is in the public domain already. The Atomic Energy Commission does not do enough to control weapons-grade material. The problem in a few years will be huge, so it must be talked about now. Describing this publicly makes hoaxes more likely, but it could lead to making real threats less likely if the A.E.C. and the International Atomic Energy Agency were to do certain things. That's the rationale by which I decided to mention this publicly, despite the obvious dangers involved. Internal pressure has not brought the needed results. The A.E.C. is up to its ears in environmentalists—in reactor-safety problems—and thus there is insufficient budget for and insufficient attention to safeguards. The United States spends billions every year preventing nuclear war; several hundred million a year is spent keeping nuclear power plants 'safe' (against accidents); not much is spent (four to six million a year) keeping nuclear materials out of the wrong hands. The actual probabilities are reversed. Nuclear war is the least likely eventuality, reactor accidents are more probable, and, by a big jump, the clandestine manufacture of a nuclear bomb is the most likely eventuality of the three. If I were convinced there was no better way to protect nuclear material than there is now, I would be slam-bang for stopping the development of the industry."

Between travels with him, I went by myself to Brookhaven National Laboratory, which is out in the rural part of Long Island, and to Charlottesville, Virginia. I wanted to ask Professor Willrich and certain physicists of Brookhaven how they would defend the idea of reviewing in public such a volatile matter as the home manufacture of a nuclear weapon. Willrich said (in capsule form), "The A.E.C. is in a state of flux and reorganization. There have been occasions when as much as a hundred kilograms of high-enriched uranium—in a fuel-fabrication plant—have been missing and unaccounted for. Meanwhile, the entire United States nuclear industry is getting clobbered by environmentalists. The investment is great. The cost of reactors has tripled. The utility companies are not making money. Then we come in and try to say that we are worried about fissile material flowing around, and they say, 'Jesus, don't bother us with that.'"

William Higinbotham, at Brookhaven, said, "Good safeguards against the theft of nuclear materials will involve a conflict between the Atomic Energy Commission and industry. We're worried that the A.E.C. will become captured, like other regulatory agencies. We'll all become victims of the industry we're trying to control. The A.E.C. is almost seven thousand people, mostly looking after dollars and cents, and looking out for their own jobs. You keep a low profile, keep your secretary busy, and look out for yourself. Eighty-one people are in safeguards applied to the nuclear-power industry. They are not always heard. The A.E.C. is overwhelmed with reactor-safety problems. The safeguards people get little support from upstairs. The International Atomic Energy Agency will be coming in to inspect the American nuclear industry. Things will erupt. The effectiveness of the International Atomic Energy Agency depends on national safeguards systems."

Higinbotham sat at a conference-room table with two other physicists. One of them, whose name was Raymond Parsick, said, "I'm pretty much of a cynic on how well a regulatory agency can control industry. A safeguards system has to be designed for maximum benefit to the public, and I think this requires public knowledge. The public has first to know what is needed and then to demand it. If you are worried about presenting a blueprint to the wrong person, that problem is least critical right now. The threat is not so great now as it will be. The future plutonium cycle is what matters, when the problem really grows. The weakness right now is on the high-enriched-uranium side."

Sylvester Suda, Higinbotham's other colleague, said, "What we need to do is to build a safeguards system for the future. We are at the crossroads right now, and the question is: Will regulations suit the nuclear industry or protect the American public?"

Parsick said, "It will be hard to get industry to cooperate. For example, if you inspect a plant every thirty days, then you have to shut the plant down every thirty days, and you lose money against foreign competition. Industry will carry a lot of clout with this argument."

Higinbotham said, "As it is now, when the A.E.C. says it has seven thousand people keeping track of material on all sides, the truth is that they only know where it is once a year. Even then, they don't know enough. When facilities that fabricate and reprocess U-235 and plutonium are inspected, the inspectors go through sets of books to see if they more or less balance. The inspectors count cans. They pull samples. They can't afford to do enough. In the end, they make a complicated statistical analysis, scratch their heads, and say, 'O.K.'—or they file a statement of lack of compliance, and *this* has to be good enough to stand up in court."

"The present safeguards system is not a system for the future," Suda said. "In order to prove suitable, it has to shape the developing industry, rather than play catch-up all the time."

What, then, if someone did have a few flasks of uranium hexafluoride? Fully enriched. Took it off a truck outside a McDonald's in Wheeling, West Virginia. Took it out of a freight room in a New York airport. It does not matter where or how the material was obtained, whether the theft was a hit or an inside job. It is hardly arguable that the material is there for the taking. If Ted Taylor, imagining himself to be the thief, had enough uranium safely sequestered, what would he do with it to convert it to a form that could be used in a bomb?

In rural Maryland, no more than thirty miles from Washington, a friend of mine has about a hundred acres of land with a cabin in the center of it. A stream runs past the cabin, and hillsides that are covered with deep deciduous forests rise away on every side. The cabin has a big fireplace, no electricity, kerosene lanterns, and a roof that projects six or eight feet over a front porch, on which there is a table and some chairs. Taylor and I went there almost every day for a week or so, sat on the porch, and looked across the stream and meadow into the woods. The place was convenient. It was near his home. He had a pair of binoculars, with which he followed birds, and a slide rule, with which he created imaginary weapons. I had notebooks and pencils, the table to write on, and a lot of leisure time, because he spoke slowly, if at all, making sure that everything he said was in a context as available to the world in public print as it was to him from memory. Nothing he said there crossed barriers of secrecy that had not already been taken down. He was pursuing, in its many possible forms, the unclassified atomic bomb.

There would be a scale of convenience. It would be much simpler to use "broken buttons"—chunks of metallic uranium-235—than uranium oxide, for example. It would be easier to begin with the right form of uranium oxide than with uranium hexafluoride. Concomitantly, though, a clandestine bombmaker would have to settle for what he could get, on a scale of availability, and he could use uranium-235 in almost any form. There is no absolute need to have uranium metal. If the oxide were used, the sacrifice in yield would not be prohibitive. The oxide is a powder, easy to handle, easy to pour. It could be packed into a box.

I asked if there would not be a density problem in using a material so relatively fluffy compared to metal.

He said, "Any high explosive that you have in the thing will see to it that the density problem disappears." The more he thought about it, he said, the more convinced he had become that the oxide would be particularly serviceable for a crude bomb, and convenient as well, for great amounts of U-235 in oxide form move around the country.

I asked him what someone would do who wanted to change the oxide into metal.

Taylor said he would put about four and a half kilograms of the powder on a vibrating tray in a laboratory furnace, and then heat up some hydrofluoric acid in a stoppered flask. Through a tube in the top of the flask, hydrogen-fluoride gas would move into the furnace. Heat the furnace up to five hundred degrees centigrade. The hydrogen fluoride and the uranium-oxide powder form water and uranium tetrafluoride, also a powder. In a ratio of six to one, put uranium tetrafluoride and powdered magnesium into a graphite crucible. Add potassium chlorate as a chemical heat generator. Put the crucible into a strong steel container. Using electrical ignition wire—like the wire in a toaster—get the temperature of the material in the crucible up to six hundred degrees. At that temperature, the uranium tetrafluoride and the powdered magnesium ignite. In combustion, they become uranium metal and magnesium fluoride. Let cool to a hundred degrees. Now spray water on the crucible and bring it to room temperature. The metal inside is known as a derby. Four kilograms of U-235. Repeat the process.

For the various procedures involved in converting uranium from one compound to another, or for bringing it ultimately to metallic form, the necessary equipment can be made at home, or can be sought in the Yellow Pages of the telephone directory, under "Laboratory Equipment & Supplies," or, for that matter, under "Hardware—Retail." Most useful of all would be the catalogue of a large chemical-supply house, such as Fisher Scientific, which has branches in most cities. The furnace costs less than a hundred dollars. A graphite crucible costs three dollars. Hydrofluoric acid costs seven dollars a quart, and magnesium oxide costs twenty-one dollars a pound. A vibrating tray is simple to make. It works with a little motor and vibrates like a bed in a Holiday Inn. Uranium oxide on a vibrating tray will mix more readily with hydrogen fluoride to form uranium tetrafluoride and water.

In 1969, Vincent D'Amico, a safeguards specialist at the Atomic Energy Commission, got word that an air shipment of fifteen kilograms of uranium hexafluoride, in a steel cylinder, was missing. He went out to search the country for it, and eventually found it in a freight room at Logan Airport, in Boston. UF_6 is the most abundant form in which fully enriched uranium travels. It comes out of Portsmouth, Ohio, in steel bottles and is distributed to conversion plants that change it to oxide or metal.

"How would you—if you had stolen some—turn UF_6 into metal?"

"Mix it with carbon tetrachloride in an evacuated nickel container. Four parts of carbon tetrachloride to one part UF_6. Heat the mixture—a stove will do—to a hundred and fifty degrees centigrade. The contents react and form uranium tetrafluoride and fluorinated carbon chloride. The UF_4 is a loose cake of solid material. Wash it with weak acid or alcohol. From there, it's the same as it was with the conversion of uranium oxide. Add powdered magnesium to the UF_4, burn it, and you get a derby of uranium metal."

The fuel plates that run certain research and test reactors are thin strips of metal only about two feet long and four inches wide. What could someone do with a stack of those?

Put them in an aqueous solution of lye and fertilizer. The lye would have to be quite pure, though—good sodium hydroxide. The fertilizer would be sodium nitrate. The fuel plates consist of an aluminum-uranium alloy sandwiched between layers of uncomplicated aluminum. After five hours in the lye and the fertilizer, the aluminum has dissolved and the uranium is in suspension. Add barium nitrate to keep the uranium from dissolving. Then put the whole business into a centrifuge—say, a six-hundred-dollar centrifuge from Fisher Scientific. Whirl it there for twenty minutes at eight hundred Gs. Pour off the aluminum solution. In the bottom of the centrifuge tubes is solid uranium-235.

"Uranium-zirconium hydride is the fuel for about half the research reactors," Ted continued. "And uranium-zirconium alloy is a step along the way to making it. The alloy is stockpiled in significant amounts, depending on business. Fuel for TRIGA reactors, for example, is made in San Diego and then shipped all over the world. If you wanted the pure alloy, you would have to steal it in San Diego. If you want the hydride, go to Bandung, or wherever, or get it in transit. The core of a standard TRIGA contains only two to six kilos of uranium, so a bombmaker would probably have to collect the stuff. If you were stealing fuel being shipped, you would have to perform at least four thefts. The reactors use cylindrical rods of hydride, clad in aluminum or stainless steel. You burn off the hydrogen at a thousand degrees, and dissolve the zirconium in sodium hydroxide."

"How much uranium is the least you might need?" I asked him.

"The classical figure for the critical mass is twenty kilograms of fully enriched uranium," he said. "The classical statement is that it takes that much to make a bomb. That statement isn't true. It takes much less—and how much less depends on how good you are at making bombs."

"How much less?"

"All I can say is it's not a nit-pick. It isn't a matter of saying twenty and meaning eighteen. It matters a lot how much less, but that is classified, and there is nothing we can do about that, I guess. But if someone gets hold of the Los Alamos critical-mass summaries, he can see how much material is critical in various forms—various ways of shaping the metal, various reflectors wrapped around it. You write to the National

Technical Information Service, in Washington, for the critical-mass summaries. They cost three dollars. In one of them it says that the critical mass varies inversely with the square of the density of the metal and reflector. If both the reflector and the core are compressed by the same amount—remember, this is an implosion system—the critical mass is reduced by the square of that amount. This is as close as you can sidle up to this classified point."

Dusk had long since come down. We quit for the day. The corners of Taylor's mouth turned down for a moment, and he said, "A small group has not had the opportunity before to rearrange people and buildings this way."

Not long thereafter, when we were in Los Alamos, Carson Mark talked about clandestine bombmaking, and he said, "Everybody has it in mind that it would be impossible to do. They say you would need your own Manhattan Project. They speak of the scale of ingenuity, of the required genius; they think of it as a tremendous operation. But the context has changed. It would not be impossible now. It does not take a fleet of Einsteins to accomplish, or even Ted Taylors, for that matter. It is not beyond reach. It is much within reach. There's a great difference between 1942 and now." Mark went on to explain that the people in Project Y (the Los Alamos part of the Manhattan Project) had faced eight principal requirements in 1942. They needed nuclear and neutronic data—energy estimates, and so forth. They needed equation-of-state data to estimate assemblies or explosions. They needed to know the probability of initiating a neutron chain. They needed a way to estimate the dependence of efficiency on various parameters—such as the mass of material, energy generation, and features of disassembly—or it would be impossible to decide if, say, five critical masses were needed for an effective bomb, or one and one-tenth, or whatever. They needed to develop numerical techniques for making neutron multiplications. They needed hydrodynamic calculations. They needed computing equipment. "And, finally," Mark said, "they needed people who could ask the right questions and suggest the significance of the answers when they found them—call them physicists, if you want. When the United States began work, it was well equipped on Item Eight—the people who could ask the questions—and on nothing else. Those people were the constellation of Los Alamos. That is what is assumed is needed now—but it is not so. You now *have* Items One to Seven. You don't need to ask the questions. You need an ingenious fellow, perhaps, but not really all that much so. He is hitchhiking on the talents of others. You don't need a lab anymore to measure cross-sections. They're all measured and published. If you need equation-of-state data, you can go over to the high school and find out what it is. Everything is unclassified except plutonium. But equations of state for heavy elements tend to be identical. See the *Rare Metals*

Handbook. Any reactor-theory textbook now will tell you the probability of initiating a neutron chain. The work that has been done on maximum-credible reactor accidents will tell you what you need to know about efficiency. You can get neutron calculations by mail. For hydrodynamic calculations, read Richtmyer and von Neumann on how to avoid the discontinuity of shocks. As for adequate computers, most airline offices have them. When people first began work here at Los Alamos, they needed to assure themselves that no unsuspected factor would vitiate the whole thing. We needed that information. It was possible, and remained possible until an explosion occurred, that something unexpected would affect the outcome. No one need worry about that now. Once you had made the explosion, you knew the unexpected wasn't there. You didn't have to comb the woods—that is the most important thing in all of this. Project Y was analogous to a group of alpinists finding the route up a mountain and reaching the top for the first time. After that, others could follow. Although you still need equipment—for example, a casting furnace—the steps you take are well prescribed. So far as we know, everybody in the world who has tried to make a nuclear explosion since 1945 has succeeded on the first try."

THE safeguarding of weapons-grade material is not widely thought of as the most significant or pressing problem that besets the nuclear industry. The safety questions involved in disposing of radioactive fission products have been reviewed more prominently, for example, as have been the day-to-day effects of nuclear plants on their immediate environments, not to mention the predicted results of a major reactor accident. A very bad nuclear-power-plant accident could kill tens of thousands of people. The plants are so carefully engineered, however, that the possibility seems remote. The reactor core would have to melt down, and the outer containment shell would have to be breached. Then an invisible, odorless, imperceptible cloud would drift downwind. How many people died would depend on how many people were downwind.

In West Valley, New York, high-level radioactive wastes from used reactor fuel—ruthenium, cesium, strontium—are buried in liquid form in steel containers set in concrete in the ground. The earth is back-filled, and markers very much like tombstones record what is buried below. In Morris, Illinois, high-level wastes are stored in a concrete basin lined with stainless steel. Thirty-six feet of water cool and shield the wastes, which are in solid form. The radioactivity of the waste material is lethal for thousands of years.

Perspective is where you find it, and in the physics laboratories of universities, in the various companies that service the fuel cycle, in the utility companies, in the Atomic Energy Commission the talk may eventually get around to safeguards, but it is more likely to dwell on other problems.

"This poor industry has taken its lumps—first from the memory of the bomb, then about radioactivity, then about thermal pollution."

"In the mid-nineteen-sixties, it was reactor safety—in demonstration plants, in test plants, in small power plants. In 1970, the environmental roof caved in. The Atomic Energy Commission was voted 'Environmental Rapist of the Year.' Radiation effects, thermal pollution became big issues. In 1971, radiation questions were beginning to be answered. Radiation is a phony issue in the normal operation of plants. The thermal-pollution matter was simply unfair. The whole electric-power industry has a thermal-pollution problem. So the issue shifted back to reactor safety. What happens when the cooling system fails and the reactor melts?"

"Nuclear power plants are redundantly engineered. They are safe."

"There is no choice, anyway. We need the electricity, wherever we can get it."

"There are problems in technology, but the real problem is in the bedroom."

"Fission products should be loaded onto rockets and shot into the sun."

"No. They are potentially too valuable—for medical purposes, and so on. Gamma rays. They should be put into high earth orbit, and saved there."

"They could be used here on earth as a heat source."

"Yes, but using fission products as a heat source is like using very old brandy to light a fire."

"This reprocessing plant is designed to contain ten years of waste storage. There is a total of two and a half cubic feet of waste per metric ton of fuel processed—three hundred tons a year. Three thousand times two and a half equals seventy-five hundred cubic feet in ten years, or, let's see, a twenty-foot cube. That's all. And nothing goes into the ground here in Illinois. Everything will remain in the tank. There is no waste-storage problem here."

"What happens to the steel drums in the ground in New York?"

"They rust through."

"The burial method is not sophisticated, true. But the soil is clay. It's water-impermeable."

"You find in this industry that the simpler things are, the better off you are in the long run. We believe in the fulcrum, and we believe in the inclined plane. The wheel we're not sure of."

"Safety was last year's problem. The next thing that is going to clobber us is safeguards."

"In the face of the very, very quick development of this technological system over the past twenty years, we should know more about the behavior of man. We have, maybe, to study, in this system, the personal interactions. We sometimes disregard the problem of human behavior."

"I can't help but believe that if something serious were to happen, if plutonium in quantity were stolen, the A.E.C. would immediately activate a plan of action that we know nothing about. I'll admit I base this on blind faith."

"I feel the A.E.C. recognizes the problem and is working on it all the time. I'm not the one to say if they're working on it hard enough."

"My gut feeling is that it's not being taken as seriously as I'd take it. It would not be my No. 1 consideration among nuclear problems, but I'd give it serious consideration."

"The commissioners sweep safeguards under the rug, and, for all their new regulations, they will continue to sweep safeguards under the rug until they have an incident."

"Safeguards can go beyond the feasibility of what you consider normal economics. You can't have Fort Knox with a cavalry division sitting on top of it. A 'foolproof' system would only minimize the odds. We have locks, watchmen, gates, alarm systems, communication with authorities. Do you want to try to steal material here? Steal it from a shipment. That is where it is most vulnerable."

"Our employees could never steal the stuff, because of in-plant procedures."

"We are assuming that the managers are honest. If management is crooked, that is a new ballgame."

"We've got the best God-damned system here that is

economically feasible. If you're all the time shutting down for inspection, you're never going to get any product."

"How did we ever get off the subject of the nice, clean, productive atom?"

"It's got a double head, and one head has to be cut off."

"If someone had written an environmental-impact statement before the first sustained chain reaction, they might not have nuclear energy to kick around today."

"Looking at the nuclear industry is like looking at a healthy-looking man who has come down with a fever. Is it a cold he is getting, or is it pneumonia?"

"The A.E.C. blew the environmental issue. The A.E.C. blew the safety issue. The A.E.C. will now blow the safeguards issue."

"It is ethically indefensible to leave future generations so much dangerous material."

"I guess all this is just one of the things that man and civilization are going to have to live with. I've heard it said it's no worse than fire."

ONE way to bypass any civilian safeguards system would be to go straight to the military and steal a bomb. That would take care of almost everything in one act. The purpose of safeguarding fissile material in the civilian nuclear-power fuel cycle is to prevent a person, group, or nation from misappropriating the material and making a bomb. All the instruments for counting atoms and all the techniques for physically protecting uranium-235 and plutonium-239 have been developed to this end. But why go to all the trouble to steal, say, uranium hexafluoride and convert it to metal and fabricate a crude weapon when the military has tens of thousands of extremely well-made bombs distributed all over the world? Some people shrug their shoulders about materials safeguards—what's the point? why bother?—in the light of the more straightforward alternative of stealing a bomb. They suggest that such thefts may well have occurred already. They ask questions like this one: "I'd be interested to know why, at any given time, the military thinks they still have all the bombs they thought they had." In 1971, Representative Craig Hosmer, of California, a member of the Joint Committee on Atomic Energy, told a safeguards forum at Kansas State University that he had once worked in military-weapons laboratories, and had also ridden the perimeters of sites where bombs were stored, and had since reviewed the procedures by which bombs are protected, and that, quite frankly, the job "ought to be done better." It has also been pointed out that in order to make a blackmail threat completely credible, if that were the purpose at hand, a home-made-bomb maker would have to make two bombs.

After one had been exploded—in, say, a desert—the bombmaker would be believed. But if a bomb was overtly stolen from the military, no one would doubt that it existed and the threat could the more simply proceed. As one engineer has put it, "If he took one, I'd say he had it."

The plutonium for military weapons is machined and otherwise prepared for assembly by Dow Chemical at a plant in Rocky Flats, Colorado—a place full of gleaming machines, where people go around in white hats, white protective clothing. Rocky Flats has been described as a combination of an automobile-assembly plant and a hospital. Fully-enriched-uranium-bomb parts are made in A.E.C. shops at Oak Ridge, Tennessee. From Oak Ridge and Rocky Flats, the uranium and plutonium go to a plant called Pantex, near Amarillo, Texas, or to another A.E.C.-contractor plant near Burlington, Iowa, where they are combined with other components in the building of nuclear bombs. Safeguards procedures for government bombs are not publicized. The bombs leave Amarillo and Burlington under guard and travel in so-called courier vehicles, by air or rail or road. Guards go with them. The guards have recently been ordered by the A.E.C. to "shoot to kill" anyone trying to steal bombs. Few would deny, though, that a hijacking effort of high sophistication could result in a successful theft.

Nuclear bombs become obsolete. Certain types become run down and diminish in potency with age. So, from all the many hundreds of stationary sites where they are kept in readiness around the world, and from all the submarines and planes that carry them, they have to be sent back from time to time to Pantex or the Burlington plant, to be disassembled, and then back to Oak Ridge or Rocky Flats to be recycled. Revitalized, modernized, they go back to Turkey or New Jersey, or wherever they belong. One would not need a crane, or even an assistant, in order to carry off a nuclear bomb. A suitcase would do.

I once visited the National Atomic Museum, in Albuquerque, with Ted Taylor. It had been open to the public since 1969, but he had never been there, and as we walked through the door he drew a shallow breath in sudden surprise. It was a hall of weapons, a long, high-ceilinged room filled with fission and fusion bombs. They were hanging from the ceiling and sitting on the floor—forty or fifty in all. A bomb called Mark 61 was dangling from a drogue parachute. It was a cylinder about twelve feet long, and it had a diameter of one foot. A plaque near it said, "Yield of the Mark 61 is in the megaton range."

"Just looking at it tells you a great deal about the design of an H-bomb," Taylor said. "Wow. This place is really something."

The museum was organized more or less along lines of chronology, beginning with Little Boy and Fat Man, the Hiroshima and Nagasaki bombs, and proceeding clockwise around the room down through the history of fission bombs to some of Taylor's minutest devices. Thermonuclear bombs, for the most part, came after that. Of course, all the bombs lacked explosive material. What was on display was the outer jackets or, in specific instances like Fat Man, Little Boy, and Mike, precise replicas of the outer jackets. That was display enough. Little Boy was ten feet six inches long and only twenty-nine inches in diameter at its widest point. It had a square, boxlike set of fins on its tail, and a blunt nose. Twelve spikes protruded from it. It was painted black. Fat Man was black, too. Fat Man had the exact proportions of an egg, with a huge box fin attached to one end. Fin and all, Fat Man was eleven feet long. The maximum diameter of the egg was five feet. On a wall near Fat Man and Little Boy were pictures of the triumphant crews that flew them to Japan—youths in coveralls, laughing, lying on bunks reading newspapers about Hiroshima and Nagasaki. In one photograph, the two pilots—Charles Sweeney, of Nagasaki, and Paul Tibbets, of Hiroshima—were shaking hands.

Taylor said that Fat Man and Little Boy were so big—and so "Model T-ish," as he put it—that no one could possibly mind showing their dimensions, but as we walked along the bombs became smaller and smaller, until he remarked that he was "not really shocked but quite surprised" to see them there. Mark 7—thirty inches in diameter. Mark 12—twenty-two inches. They were ten or twelve feet long and were shaped like sharks, but they were built to contain spherical fission bombs. Davy Crockett was the end of the line—a small bomb that had the exact proportions of an egg, with a box fin attached to one end. The egg part was just about the size of a Rugby football. The whole bomb was about two feet long, and its maximum diameter was twelve inches. Taylor had designed it. "That is what happened to Fat Man," he said. "You can see why I used to get so riled up, in my first year at Los Alamos, when people said, 'With fission bombs, there's nothing left to do.' "

One display was a cross-sectional mockup of the Terrier missile, half scale, and the dimensions of its warhead cavity indicated that the bomb the Terriers carry could not have a diameter greater than ten inches—probably a near cousin of Davy Crockett. Mark 48 was an artillery shell about three feet long, with a diameter of six inches. That strongly suggested a gun-type device within, and I asked Taylor if that's what it was, and he said he couldn't be sure. A complete implosion bomb could, in other words, have the diameter of a small cantaloupe. In the nose of another missile he pointed to an area about the size of an orange, and he said, "I'll tell you right now, you can't put an A-bomb in there. Not quite."

Mike, the first hydrogen bomb, sat at the far end of the room—a brooding presence, out of place. In a hall of tapered, aerodynamic projectiles, Mike was just a huge, five-hundred-cubic-foot hot-water tank, lying on its side—a big cylinder, with no taper, twenty-two feet long. Its progeny, one after another, and progressively smaller, were set in a row beside it—cylinder after cylinder, no taper. "The shape tells you a lot about H-bomb design," Taylor said again. "But not enough." I drew a sketch of a hydrogen bomb showing a cylinder full of thermonuclear fuel, with *two* fission bombs, one at each end, so built that their explosive force would travel primarily through the thermonuclear fuel and meet in the middle. Looking over this pathetic effort, he said, "Nice try, but that is not what happens."

He was resting his arm on an H-bomb called Mark 41, labelled only, like the others, with its name and the phrase "in the megaton range." It was just a tube a few feet long, like a section of pipe. I asked him how much it could destroy.

"A great deal," he said. "But not as much as people sometimes think. People imagine doomsday, but even an all-out nuclear war would not bring doomsday. Remember, the largest bomb ever exploded was sixty megatons. Mike was ten. This one is a great deal less than that. If one wanted to, I suppose, one could imagine a single superbomb that would kill all things on earth, that would deliver a hundred thousand roentgens to every part of the world and leave life belowground ecologically strangled. Such a bomb would be a quarter of a mile long, though, and five hundred feet high. It would probably have to be deeply buried to keep from blowing off the top of the atmosphere and dissipating too much energy into space, but not so deeply buried that the explosion would be contained. It would make a crater perhaps hundreds of miles in diameter. The yield would be many millions of megatons." On a wall of the museum, near the door, was a picture of two young men with their sleeves rolled up lifting the core of the first bomb out of a 1942 Plymouth, which had carried the device the two hundred miles from Los Alamos to Alamogordo in 1945. It was not much of a load—less than twenty pounds in all. Beside the picture was a framed copy of the front page of the Albuquerque *Tribune* of July 16, 1945. The headline said, "MUNITIONS EXPLODE AT ALAMO DUMP." The story began, "An ammunition magazine exploded early today in a remote area of the Alamogordo Air Base reservation, producing a brilliant flash and blast, which were reported to have been observed as far away as Gallup, two hundred and thirty-five miles northwest."

The symbol of Project Y was the letter Y framed in a circle—an inverted peace sign. A sample of the wax seal by which this symbol was affixed to secret documents is on display in the Science Hall and Museum at Los Alamos, which, like the National Atomic Museum, in Albuquerque, is open to the public. The Los Alamos museum is the more subtle. It includes the skull of a twenty-year-old Indian female who died near Los Alamos around 1580. It includes pictographs, petroglyphs, Indian bowls, tools, axeheads, and the points, or warheads, of obsidian projectiles. It includes Oppenheimer's chair, and a formal requisition in which the United States government was asked to provide a nail on which Oppenheimer could hang his hat. It includes a melted windowpane from Nagasaki and a glasslike rock that was once sand beneath the tower near Alamogordo. I remember Stan Ulam in that museum. He had been wandering along talking to Taylor, and they happened into the museum. Ulam went across a room and opened a door to an interior courtyard. Four bombs were there: a 1961 thermonuclear bomb, eleven feet long; a 1962 gun-type fission bomb, eight feet long; and Little Boy, of Hiroshima, and Fat Man, of Nagasaki. They were not black, as in Albuquerque. They were all painted pure white, up here in Los Alamos, and were dazzling in the light of the sun. Ulam, holding the door open, said, "See there. That is the Grand Guignol."

LUNCHTIME on a bright fall day at the Maryland cabin, Taylor opened a can of beer, took a bite of a ham sandwich, and began to consider what might be done with plutonium. Some people thought plutonium was too difficult to handle and would only frustrate an amateur bombmaker, but he did not agree. Seaborg, a co-discoverer of plutonium, had said, "It would take sophisticated, scientifically literate gangsterism to cope with it," and Taylor had no quarrel with that. True, uranium would be easier to deal with, but, for various reasons, he felt that plutonium was the material more likely to be used in a clandestine bomb. In the first place, a designer of nuclear weapons would select plutonium over uranium in the same way that a cabinetmaker would choose mahogany over yellow pine. The mass of plutonium needed for a sustained chain reaction is a third that of uranium. Plutonium's efficiency in an explosion could be expected to be far greater. A bad bomb made with plutonium might produce more yield than a fair one made with uranium. Moreover, with the coming of breeder reactors, and with plutonium recycle, plutonium will be, for thievery, by far the more available material.

The process might begin with plutonium nitrate in solution, because that is the form in which plutonium has generally been shipped and stored. In each four-foot flask, which is like a long thermos bottle, is about two and a half kilograms of plutonium, and, even if the bombmaker were not very skilled, only three or four flasks would be enough. Shipments include twenty to thirty flasks. Taylor guessed that a relative amateur, proceeding cautiously, would probably refer to the *Plutonium Handbook* (Gordon & Breach, New York, 1967, two volumes; $81.50), a guide to plutonium technology, and to the *Reactor Handbook* (Wiley, New York, 1960–64, four volumes; $123), which he called a how-to book that contains details of plutonium processing. Plutonium would require equipment on a scale more complicated and expensive than would uranium. The most expensive item needed might be a fifteen-hundred-dollar induction furnace, a device that produces a magnetic field and can heat up to high temperatures anything within it that is resistant to electricity—for example, plutonium. Such furnaces are sold by metallurgical-equipment-and-supply companies. They are also available at Fisher Scientific. A crucible in the ten-dollar range would be needed, too—and, of course, a glove box. The Stainless Equipment Company, in Denver, sells glove boxes for a few hundred dollars apiece, but a wooden one, made at home, would do well enough.

In one particular sense, Taylor said, a person trying to make a plutonium bomb at home today would indeed be imitating the Manhattan Project. Carpentered wooden glove boxes were used at Los Alamos in 1945. The first bomb, the one exploded near Alamogordo, was a plutonium bomb, and it was made at Los Alamos in an old icehouse. The making of the bomb itself was only the last and least onerous of the many tasks of Project Y. Chemists and metallurgists scavenged the country for supportive equipment. Eventually, they got the bomb together in circumstances not importantly different from what someone might do in a private way these days. Taylor said he remembered his friend Dick Baker, a metallurgist, telling him that he had worked weekends fashioning ceramic crucibles with his own hands in preparation for the making of the Alamogordo bomb. (Baker, amplifying the story himself, would tell us at Los Alamos, "Frankly, the only unique thing to the production of an atomic weapon is the fissile material. We developed no great special equipment for the bomb during the war. We just put fissile material with commercial materials. It's not that complicated, if you see what I mean. The early bomb work was something like what might happen in a garage now. For the reflector and so forth, all you need, frankly, is a good machine shop." A slight and gentle person, around sixty, in rimless bifocal glasses, Baker went on to say, "Trained people could work plutonium without getting into a serious health problem. But in order to perform a clandestine operation you don't need to be as conscious of safety as we are at Los Alamos. People are sort of expendable, you know. You could have a bomb that didn't have to be near as refined as the first ones we made here and you'd still have a bomb.")

Taylor mused on, ignoring his beer and his sandwich. This was October in Maryland, not wartime in Los Alamos, but what might happen in a secluded place like this was pretty much what had happened there. The typical way to get metal out of a solution is to add something that makes an insoluble compound with metal. When the compound precipitates, filter it. Add, say, oxalic acid to plutonium nitrate. The precipitate is plutonium oxalate in crystals that hold water. Remove the water with heat. Now you have a cake of anhydrous plutonium oxalate. Now further dry it by running a stream of hydrogen-fluoride gas through a sealed crucible.

"How?"

"Oh, simply enough. Buy some hydrofluoric acid and heat it in a quartz container—ten dollars. Hydrogen-fluoride gas comes off and goes into the crucible through a quartz tube. Anhydrous plutonium oxalate cooked at five hundred degrees centigrade in hydrogen fluoride becomes plutonium fluoride. Do it in batches of a few hundred grams—small enough to avoid going critical. Build up a stockpile of plutonium fluoride. Now line a crucible with magnesium oxide. You mix it with water and make a paste. Form it. Work it. Dry it. It's like clay. This is what Dick Baker did at Los Alamos in 1945. Now get some metallic calcium and crystalline iodine from a chemical-supply house. Put five hundred grams of plutonium fluoride in the crucible. Add a hundred and seventy grams of calcium and fifty grams of iodine. Cover with argon, which is a heavy inert gas. Close the lid. Now heat up the crucible inside an induction furnace to seven hundred and fifty degrees centigrade. At that point, the mixture in the crucible reacts and, in one minute, heats *itself* up to sixteen hundred degrees. The pressure is considerable. In the next ten minutes, the whole thing cools itself to eight hundred degrees. Remember, this is just what they did to make the first bomb. Now remove the crucible from the induction furnace. Let it stand until it comes to room temperature. Open the lid. Metallic plutonium is in there with some calcium-iodine junk. Use nitric acid to wash off the iodine flakes and the calcium-fluoride salt. What you have left is a small lump of plutonium. You can hold it in your hand. It won't hurt you. It feels a little warm, from alpha decay."

Taylor fell silent for a while, and there was no sound but some wind in the turning leaves. Finally, he said, "Of course, you could just boil away the water, get plutonium-nitrate crystals, and make a bomb out of the crystals. It would not be much of a bomb—only a tenth of a kiloton, say—but that's enough to knock down the World Trade Center. When recycle comes in, the reactor-fuel rods will probably contain mixed uranium and plutonium oxides. The uranium would be only slightly enriched and not usable in a bomb, but the plutonium would be. There is good, better, and worse, but there is no nonweapons-grade plutonium involved in the nuclear industry. Mixed uranium and plutonium oxides can be separated chemically. It's not a difficult thing to do, but you'd better be pretty careful, because plutonium is so poisonous. You would add nitric acid to put the pellets into solution. Then add oxalic acid. The plutonium forms plutonium oxalate. The uranium remains in solution, because it does not combine with oxalic acid. So you filter out the plutonium oxalate with filter paper. Now you are where you were when you had plutonium oxalate before—and you proceed to make metal. Or you can heat up the plutonium oxalate in a furnace to a thousand degrees centigrade, and you get plutonium oxide, which is weapons material in itself, although less efficient. When plutonium recycle really gets going, incidentally, this stuff—plutonium oxide, out of which bombs can be made directly, without any chemical processing—will exist in huge quantities all over the country, both in transit and in storage." The afternoon by now was well along. Taylor's beer remained open but untouched. He had taken two bites out of his sandwich.

A BOEING 707 banked in the sky above us, flashed in the sunlight, squirted hydrocarbons, and lowered its flaps for Dulles. Taylor watched it go out of sight over the hill behind the cabin. Minutes later, the raw smell of jet fuel fell into the air around us—all over the woodland and the valley stream, incongruous in place if not in time. The smell was the same as from the lantern fuel inside the cabin. Taylor seemed to be working numbers in his head. After a time, he said, "It took ten million kilowatt-hours of electricity to make that airplane. It also takes a great amount of energy just to keep it up in the air. When you stop and think about it, it doesn't make any sense to fly." He said he would like to explode his way across the United States deep underground, making with nuclear bombs a tunnel from New York to San Francisco in which vacuum-tube trains would move as fast as that nonsensical jet.

Such trains would be cheap to run and would have no effect on the biosphere.

"Yes, and Pittsburgh, Chicago, Omaha, and Denver would fall into your holes in the ground," I said.

"Oh, my goodness, no," he said. "You don't just gouge out one cavern after another in the rock. You shape the charge."

He paused, and his gaze went into the middle distance and rested on a meander in the brook. "Suppose a deer were standing over there and you were hunting it," he went on. "You could kill it with a bomb that weighed many pounds. Or you could kill it with a bullet that weighed a few grams, if you knew how to direct the bullet. A nuclear explosion is not completely symmetric. Its asymmetries can be enhanced by rearranging the way the thing is put together. Various kinds of energy come out in the explosion. You can, if you want to, concentrate energy of a particular sort in a particular direction. You can, in effect, fire it like a bullet.

"You could perform an experiment with TNT that would show you just what I mean. Set five pounds of TNT next to a huge block of steel. Explode the TNT. It will make a quarter-inch dent, maybe not even that much, in the steel. Now take another five pounds of TNT and form it into the shape of a C about six inches high. Line the inside of the C with tungsten, or any dense metal, and set the thing back about three feet with the opening in the C facing the steel. Detonate it. A projectile of liquid tungsten will cut a hole half an inch in diameter two feet into the steel. It would go fifteen feet into a wall of rock. It's exactly like a straw in a tornado being driven through a telephone pole or the roof of a house. A one-kiloton fission device, shaped properly, could make a hole ten feet in diameter a thousand feet into solid rock. And the reason such a tunnel hasn't been dug by now is that the American engineering community is essentially conservative as hell.

"A great variety of things, many forms of energy, come out of a nuclear explosion—gamma rays, alpha particles, neutrons, X rays, visible light, radio frequencies, radar frequencies. To some extent—and in all cases, to an important extent—you can select what to enhance and what to suppress. The relative amounts and directions can be controlled over very wide ranges. There are so many things you can do—through conceptual design. If you want a bomb that spews out nothing but green paint, you can do that."

Ted said he would admit to a pure fascination with nuclear explosions, a fascination wholly on an intellectual plane, disjunct from practical application. Down the years, it had been a matter of considerable anguish to him to live with the irony that what he thought was the worst invention in physical history was also the most interesting. He said he had been hopelessly drawn to the spectacular and destructive potentialities of plutonium, even from the first moment he had ever heard its name, and to the binding energy that comes out of the nucleus and goes into the fireball, even before he could come to grasp the stunning numbers that describe it.

A proton on its own, free, weighs more than it does when it is inside an atomic nucleus. The same is true of a neutron. The difference in weight is extraordinarily small—in each case, a few octillionths of a gram—but a difference is there. How could something that is indivisible, a fundamental block in the construction of matter, have more weight in one situation and less in another? Several centuries of physical science indicated that this could never be, because it was a law of physics that matter could not be annihilated but could only be changed into other forms of matter. What, then, happens to the few octillionths of a gram of a neutron or a proton when it gets inside a nucleus? Why the slight loss in weight?

A law concomitant to the law of the conservation of matter had been the law of the conservation of energy. It was believed that energy could never be lost, either—that it could only turn into other forms of energy. Relativity theory, coming around the turn of the twentieth century, suggested that these two laws were independently incomplete, that energy could in fact turn into matter, and that matter could turn into energy. What has happened to the neutrons and protons losing weight in the nucleus is that a minute part of each of them has turned from matter into energy. This energy is equivalent to the strength of the forces that bind the parts of the atomic nucleus together—hold the protons and neutrons in there as a unit. Hence, it is known as binding energy. The first measurements of the binding energies of all known elements showed a remarkable curve. From the lightest elements upward, binding energies grew generally stronger and stronger, until—somewhere in the region of cobalt, nickel, and copper—they reached a peak. Binding energies then were seen to grow weaker and weaker as the elements progressed in weight toward uranium. This graph became known as the curve of binding energy. It suggested this: If an atom of a very heavy element, with a couple of hundred protons and neutrons in its nucleus, were to be split apart, the protons and neutrons would convert mass into energy as they formed other elements higher on the curve. Uranium-235 breaking apart, for example, might become tin and molybdenum, and the uranium's two hundred and thirty-five protons and neutrons would busily create energy while giving up mass. The same sort of thing should happen at the other end of the curve if the atoms of two very light elements, instead of being pulled apart, were pushed together. If two hydrogen isotopes were pushed together to make helium, the protons and neutrons would have to shrink a little, and in so doing release energy. As Henry D. Smyth would eventually write in *Atomic Energy for Military Purposes,* "Any nuclear reaction where the particles in the resultant nuclei are more strongly bound than the particles in the initial nuclei will release energy." The amount of difference in particle weights was so nearly infinitesimal that it might be hard to imagine a great deal of energy being created in the processes described. It had been Einstein's suggestion that the energy would be equal to the mass that disappeared multiplied by the square of the velocity of light. The velocity of light is about a hundred and eighty-six thousand miles per second, and the square of that is thirty-four billion five hundred and ninety-six million. Thus, the loss of weight in a proton or a neutron might be almost nothing, but to get a sense of the amount of energy derived from the annihilation of such a small amount of matter one would multiply by an extremely large number. If the mass-energy change were to happen to many atoms all at once, the result would be an amazing explosion. A bomb might be made by disintegrating, or fissioning, uranium. A fusion bomb could conceivably be made by forcing lightweight atoms together at thermonuclear temperatures. By the time all this had been theoretically established, it was 1939, and a most sensitive question was: Who would be the first to achieve a military application of the new knowledge? Einstein felt that it was his duty to send to President Roosevelt a letter telling of the possibility of a nuclear bomb and predicting that "a single bomb of this type, carried by boat or exploded in a port, might very well destroy the whole port together with some of the surrounding territory." The letter went on to say that "such bombs might very well prove to be too heavy for transportation by air." The bomb that was exploded near Alamogordo on July 16, 1945, was called Trinity. Its core was designed by a man named Christy. The name of the exact place where it was detonated was Jornada del Muerto, the Journey of Death.

"My memory has never been very good," Taylor said, "and it is getting detectably worse. There was a time when I could remember all the numbers for every bomb, from Trinity up to the moment."

I asked how many numbers.

"Oh, the yield, the diameter, the mass of the high explosive, the mass of the reflector, the mass of the core, the initial radius of the core, and the alpha, which is one divided by the neutron-generation time."

He said that Carson Mark had once pointed out to him a number, a fact, that brought with it the most astonishing realization he had ever experienced in physics. It had to do with binding energy, and it was that when Fat Man exploded over Nagasaki the amount of matter that changed into energy and destroyed the city was one gram—a third the weight of a penny. A number of kilograms of plutonium were in the bomb, but the amount that actually released its binding energy and created the fireball was one gram. *E* (twenty kilotons) equals *m* (one gram) times the square of the speed of light.

"In physics, I am opposite to what I am in life. In physics, I like extreme situations. I don't like intermediate steps. I am attracted to the extremes: the highest-pressure places, the highest-temperature places, the greatest speeds, the greatest densities—and all these are within a nuclear explosion. I was always looking at them, because I was always trying to make as light a bomb as was possible in principle. When an implosion bomb is detonated, the temperature in the core builds up to several hundred million degrees in one hundred-millionth of a second. That is many times the temperature in the center of the sun. It's a temperature high enough to strip the electrons off any but the heaviest elements. All you're left with is the bare nucleus in a sea of electrons. (In a hydrogen bomb, the temperature can be five times as high. It strips off electrons even from uranium—almost all ninety-two electrons gone! It's incredible.) In an implosion, pressures at the center build up to a hundred million atmospheres, and with these pressures the core begins to expand at speeds of —let's see, two times ten to the eighth centimetres per second—about five million miles an hour. Meanwhile, neutrons are multiplying, with a whole new generation every hundred-millionth of a second. (At the center of an efficient thermonuclear explosion you have so many neutrons that they actually form a gas with the density of a metal.) Plutonium and uranium split unevenly. It is rare that they split into two equal parts, and in the explosion their fragments become every element below them. Anything you can name is there—molybdenum, barium, iodine, cesium, strontium, antimony, hydrogen, tin, copper, carbon, iron, silver, and gold. I am trying to describe the beginning of the explosion—something twelve inches across expanding faster than anything in our galaxy. Conditions there are quite different, perhaps, from anything else that happens in the universe, unless there are other people who make bombs."

SOMETIMES at the family dinner table, Ted Taylor will leave a conversation. He simply goes away, in every sense but the physical presence of his body in a chair. He stays away for varying lengths of time. When his thoughts have made their journey and come back, he will resume a conversation at the exact place he left it, as if all animation in the world had been suspended while he was gone. There was an entire weekend in 1957 during which he didn't come back at all. Not long before, he had moved with his family from Los Alamos to San Diego, and he was now working at a civilian laboratory that had been set up to make creative use of nuclear energy. Russia had just sent the world's first satellite into orbit, and Taylor had become occupied with the contemplation of ways to put big payloads into space cheaply—very heavy things, adequate for space exploration, not little capsules that might go into orbit for tens of millions of dollars. He concluded that the only way to do that was with nuclear energy. One of his wife's aunts was coming to visit for the weekend, and although he remained very much in and around the house, he made his metaphysical disappearance some hours before she arrived. He thought about Rover, a nuclear rocket under development at Los Alamos, but that was simply a reactor through which you pumped hydrogen. It could place on the moon a greater payload, by a factor of two or three, than a chemical rocket of the same weight at launch. But that was not enough—not for a ship of the size that he was beginning to conceive. He sought factors of at least a hundred. Eventually, he remembered Stan Ulam's idea that space payloads might be propelled with nuclear explosions. Ulam and Cornelius Everett had worked out calculations of the momentum transfer between a series of nuclear explosions and a mass. Taylor walked around and around his house, on an interior circuit through several rooms, and also around the outside. Once during the weekend, his wife had a chore to do that involved her carrying heavy buckets of sand. Her aunt said to her, "Caro, where is your husband? Why can't Ted be doing that?"

"He's thinking," Caro explained.

At first, one might imagine the idea—nuclear explosions driving spaceships—to be ridiculous. Surely an atomic bomb exploding close to a spacecraft would vaporize it then and there. Taylor remembered that on Eniwetok one time a fission bomb had been exploded from a tower and had knocked the four steel legs of the tower outward in four directions. The heat around the steel had been many times more than enough to vaporize it, but after the explosion the four struts were lying pretty much undamaged on the ground. Beforc the heat could destroy them, the shock wave had shoved the fireball up into the air.

This sort of phenomenon had to do with a field of study—weapons effects—in which Taylor had long taken a special interest. Two laboratories were busy making bombs, and no one, in his view, had been paying enough attention to various things the bombs might do.

He thought there should be a third laboratory—a national weapons-effects laboratory—set up to discover new potentialities for the use of nuclear explosives, military and otherwise. Suppose you wanted to get rid of a city's port facility with a tidal wave and not get rid of the whole city. How would you do that? How could you dig a tunnel from New York to San Francisco? How might you destroy an enemy missile in flight without doing any damage to anything else? It was all in the special arrangement of the explosive, the enhancement of certain characteristics to obtain certain effects. How could you shove something that weighed a thousand tons into space? A fission bomb expands at an extremely high speed—about two and a half million miles an hour—and since the velocity required for escaping the earth's gravity is only twenty-five thousand miles an hour, a fission bomb might do. Clearly, it should be a shaped charge. The explosion should go in only one direction, as from a nozzle. What would it strike against? Once at Eniwetok, a physicist named Lew Allen had actually conducted a test to see what would happen if spheres of steel covered with graphite, the size of big pumpkins, were dangled from wires thirty feet from a twenty-kiloton bomb. If the struts of a tower had come through undamaged, how about the graphite-covered balls? The bomb was one that Ted Taylor had designed, and it was called Viper. Its shock wave took the balls with it. When the explosion was over, the balls were integral. Their steel interiors were undamaged. A few thousandths of an inch of graphite was gone from their surfaces. Why not set a nuclear bomb under a plate of steel and graphite on the bottom of a big ogival spacecraft, detonate it, and start for Mars? A short way up, lob another bomb out of the ship's interior. Detonate it. About a second later, lob out another bomb, and so forth. The ship would go straight to Mars, cutting across gravitational fields, violating all the rules for saving energy. There would be no sneaking in the back door, going the long way, as you would have to do with a chemical rocket, to save fuel. This ship would go essentially in a straight line, with a superabundance of energy, and it would be large enough for a crew of a hundred and fifty. During the opposition of Earth and Mars, which occurs every couple of years, the distance between them can be as little as thirty-five million miles or as much as sixty-three million miles. The round trip would take from three to six months. Long after Caro's aunt had gone, Ted asked why she had never arrived.

In 1956, Frederic de Hoffmann, once of the Theoretical Division at Los Alamos, had attracted physicists, chemists, and engineers to a schoolhouse in San Diego for a series of conferences on atomic energy. The building was the temporary headquarters of General Atomic, a division of General Dynamics headed by de Hoffmann. De Hoffmann was only thirty-two, but there was a mixture in him of sound physics and entrepreneurial verve that drew people of the highest level to the schoolhouse: Hans Bethe, Glenn Seaborg, Edward Teller, Freeman Dyson. Alvin Weinberg, of Oak Ridge, was there, and so was Manson Benedict, of M.I.T. Ted Taylor was there, and Marshall Rosenbluth. Mornings, they heard lectures in nuclear technology. Afternoons, they discussed things that might be built. At one such conference, Teller made a keynote talk. He said that what the world needed was "an inherently safe reactor"—something that you could, in effect, give to schoolchildren without fear of hurting them. Three teams were formed—a safe-reactor team, a ship-reactor team, and a team that would explore theoretical possibilities for a high-temperature gas-cooled reactor. The schoolyard was equipped with picnic tables that had blackboards for tops and chalk in recessed pockets. The tables were surrounded by bougainvillea and cups of gold. Mayonnaise got all over the blackboards, and calculations tended to skid. The blackboards were not used much. Taylor, Dyson, and Andrew McReynolds, an employee of General Dynamics, were on the safe-reactor team. Taking advantage of something called "the warm-neutron effect," they invented a reactor that would go subcritical if it got too hot—and thus, in an emergency, would shut itself off. It was called TRIGA. Taylor went to work for General Atomic. Dyson went back home, to the Institute for Advanced Study, in Princeton. There are fifty-three TRIGAs now—research reactors—in fifteen countries around the world. They are absolutely free of ordinary reactor-safety problems —no meltdowns could ever occur, or poisonous clouds break out from their containment. TRIGAs could be built, in much larger form, as commercial power reactors, but they gobble up neutrons and might be too expensive to be competitive in the business world.

Taylor's spaceship required federal support, and it got nowhere coldly until April, 1958, when it was presented to Roy Johnson, chief of the Advanced Research Projects Agency, in Washington. Having looked at all sorts of paper payloads sitting on paper rockets, Johnson was suddenly confronted with a plan for a trip to Saturn, among other places, in something the size of a sixteen-story building. He said, "Everyone seems to be making plans to pile fuel on fuel on fuel to put a pea into orbit, but you seem to mean business." The agency checked out the project—Marshall Rosenbluth had pulled together what he called "the first rough crack at the feasibility physics"; de Hoffmann's General Atomic would be in over-all administrative charge; and Ted Taylor, the ship's inventor, would serve as project manager. The agency decided that it was a crazy idea from some very good people, and offered a million dollars to support it in its first year. The project was named Orion.

General Atomic assembled, eventually, about forty people to work on Orion—under maximum-security conditions. Among them was Dyson, who, when he heard about Orion, had taken a leave of absence from the Institute for Advanced Study. He moved to California and went to work full time with Taylor. The presence of Dyson was in itself an antidote to skepticism. That he would give up everything else he was doing to assist in one endeavor could not help but signify much about that endeavor. To Dyson himself, Orion suggested not only a scientific instrument but an imperative for the future of the world. He saw the human race running out of frontiers, and he considered frontiers essential to the human psyche, for without them pressures would build that would implode upon the race and destroy it. The planets were unpromising, because of their apparent inability to support life. Dyson speculated instead about comets. Comets had abundant water and, among other things, nitrogen and carbon. They seemed to be logical places to colonize. Extrapolating from the frequency with which comets come into the solar system, it could be concluded that comets by the thousands of millions must be out there in space awaiting colonists. To provide warmth and air, trees would be grown on comets. The leaves would be genetically reprogrammed to adapt to conditions of space. Nothing would inhibit growth on a comet, so the trees would reach heights as great as a hundred miles. Returning in a sense to an earlier *modus vivendi*, people would live among the roots of these great trees, whirling through space with the basic requirements for life ready to hand.

Dyson reasoned that going through a series of energy crises would be a common experience to all civilizations in the universe. After running through the resources on its own planet, a given civilization would then logically turn to the nearest sun. The minuscule fraction of total sunlight that actually strikes a planet could not be of extensive use, so a resource-impoverished civilization, in order to assure almost indefinite survival, would send giant plates of materials into orbit around its sun, forming a great discontinuous shell, a titanic nonrigid sphere, conserving almost all the heat and light and photosynthetic sustenance the sun would give. To this end, Dyson imagined, an advanced civilization could dismantle a neighboring planet—one comparable to Jupiter in our system—whose mass would supply enough material for a shell around the sun.

No chemical rocket making slow ferries to the nearby moon was ever going to hint at the vehicular capabilities necessary for enterprises on such a scale. Ted Taylor's Orion was something quite different, though. Large enough to carry machine shops and laboratories, it could move through space at about a hundred thousand miles an hour, top speed. Whenever the day might come that people would earnestly wish to get about in the solar system, this would be the way to do it.

Dyson, a professor at the institute that had been the working milieu of Einstein, had already taken a singular place on the highest level of theoretical physics in the twentieth century, so the impressions he formed of Ted Taylor in San Diego are illuminating not only of Taylor but of Dyson: "As a mathematician and physicist, Ted was slow. It took him a long time to understand things on the technical level. He is a splendid example of the man who ripens late. Ted was not able to learn a great deal from books. He is a special kind of physicist, with a feeling for something as a concrete object rather than for equations you write down about it. In a European system, after an experience such as the one he had at Berkeley, he would never have had a chance. I have a low opinion of higher education. Ted had no time for such nonsense, and in that respect he was like Einstein. He was like Einstein, too, in his style of thinking. Both were theoretical. Neither did physics experiments in the conventional sense. Both of them were extraordinarily unmathematical. Ted thinks of real things. He does not think in equations. Einstein, in his young days, was the same way. His thought processes were extremely concrete. Ted taught me everything I know about bombs. He was the man who had made bombs small and cheap. For Orion, having them small and cheap was the point. We worked together on the problem of designing the bomb units. The problem was to blow out the debris in one direction as far as you could, with a controlled-velocity distribution. Very few people have Ted's imagination. Very few people have his courage. He was ten or twenty years ahead of the rest of us. There is something tragic about his life. He was the Columbus who never got to go and discover America. I felt that he—much more than von Braun or anyone else—was the real Columbus of our days. I think he is perhaps the greatest man that I ever knew well. And he is completely unknown."

Essentially flat on the bottom, Orion was going to look like the nose of a bullet, the head of a rocket, the hat of a bishop. The diameter would be a hundred and thirty-five feet. The intended launching site was Jackass Flats, Nevada, where Orion would rest on a set of eight towers, each two hundred and fifty feet high. At the end of the countdown, it would rise into the sky on a columnar fission explosion. In Taylor's words, "It would have been the most sensational thing anyone ever saw." Inside Orion would be two thousand nuclear bombs. Stored in cans, they would be dispensed one at a time down a shaft and through a hole in the bottom of the ship. For insight into the engineering of this mechanical operation, the Coca-Cola Company was consulted with reference to the technology of its coin-operated Coke machines. Apparatus of the Coke-machine type would move the bombs out of storage bays and set them up at the head of the shaft. Then they would be blown out of the ship by compressed nitrogen and detonated about a hundred feet below. The initial launching bomb would yield only a tenth of a kiloton. The next bomb, a second later, would yield two-tenths of a kiloton. Two hundred kilotons in all would be needed for the ship to get out of the atmosphere, and this thrust would be delivered by fifty bombs of graduated range, the fiftieth of which, at twenty kilotons, would be of the force that destroyed Nagasaki. Each bomb would need something stuck to it that would become debris and go up against the ship, pushing more emphatically than would the shock wave alone. Hydrogen was best for the purpose, but water would do, and so would polyethylene. The successive explosions would fling plastic at the spaceship to make it go. The National Aeronautics and Space Administration called Orion's nuclear explosives "pulse units." The Air Force called them "charge propellant systems." Taylor called them bombs.

The pusher plate—the bottom of the ship—was its most important component, what with twenty-kiloton bombs going off a hundred feet away. Intense effort went into the consideration of substances of which the pusher plate might be made. Substances tried in experiments included steel, copper, aluminum, and wood. Ironwood might do, because wood is a poor conductor of heat. In one experiment, a couple of pounds of high explosive were detonated a foot from an aluminum plate that was backed by heavy springs. The plate shattered. It had been a uniformly thick (quarter-inch) disc. A thinner aluminum plate was made. Its thickness tapered toward its edges. Boom. It did not shatter. Uneven stresses had broken the other one. These stresses disappeared in the taper of the thinner one. A lesson learned—you taper your pusher plate. Each explosion could remove a few thousandths of a centimetre of the plate's surface but no more, since the pusher plate was the one thing in Orion that could not be replaced during a voyage. Someone left a thumbprint on an aluminum pusher plate before a test explosion. The grease of the thumbprint prevented explosive erosion altogether. A lesson learned—grease would be sprayed onto Orion's bottom between shots.

Above the pusher plate would be a set of huge pneumatic "tires" connected through a chassis to gas-filled piston shock absorbers fifty feet high. A person riding in the chambers above would experience what Taylor describes as "a pulsating effect, a bouncy ride—but not too much so." In all, there would be seventy-five tons of shielding between the passengers and the explosions. Weight was no problem. Orion could carry tractors, big telescopes, two hundred tons of water, a complete kitchen, refrigeration, washing machines, toilets. "We had an aversion to weight-minimizing. We did not need to recycle urine, for example. We would have just thrown it over the side. We could have taken barber chairs, if we wanted them. Anything could be carried that might be necessary for a big-scale manned expedition anywhere in the solar system."

Taylor lived in a one-story clapboard house on a hillside in La Jolla, with orange and tangerine trees in his yard and a view of the blue Pacific. Orion began in the schoolhouse, but de Hoffmann in time moved General Atomic to three hundred acres of country land, where he built his nuclear Xanadu of circular and curvilinear buildings surrounded with tennis courts, pools, eucalyptus, hibiscus, ground cedar, bougainvillea, and secluded stone benches for classified discussions. Ted barely noticed. He was on his way to Pluto. He meant to go himself. It never crossed his mind that he would not. He told his children that he was taking tail medicine every night, and that he would grow a tail that would help him keep his balance during stopovers on the moon. The children never saw the medicine, but they believed him, and occasionally they felt his lower spine to see how the tail was coming along. At night, he would lie down with his children on a canvas tarpaulin in the back yard and show them Orion and Mars. He described conditions on all the planets. All over the kitchen walls he put up diagrams of pusher plates, each nuance of which was explained at length to Caro.

"I don't think it ever mattered if I understood what he said or not. The farther he got into space, the more earthbound I became. There are so many things to do, and someone had to do them. We had a lot of children, for goodness' sake."

"I never imagined myself sitting at the throttle. I dreamed of looking out a porthole at the rings of Saturn, sometimes the moons of Jupiter. We would not have landed on Jupiter itself. The mass and gravity are so great that if you put something down on it it would probably sink into a soup of methane, or whatever it is. It doesn't sound like much of a place to go. The remotest place I expected to see was Pluto, where the sun is a pale disc and there is deep twilight at noon. Cold, cold, cold. But still a world."

"What he really wanted was a rock of Mars on the mantelpiece."

"People have worked out what it would take to graze the sun—to get inside a flare, so that the flare goes over your head. Imagine looking out a window and seeing that! To think of the sun filling up half your field of view is almost unbearably exciting. I had a recurrent dream that began with Orion. I was alone in the ship, and outside it I saw many stars. I busied myself within the ship and, a little later, looked out again. There was nothing. Nothing at all. I was beyond the stars, beyond the universe. I woke with a feeling of total terror. For that matter, whenever I saw myself looking through a porthole at Jupiter filling half the sky I would get the scared feeling I have had since childhood, and which I have never understood. In La Jolla, I had a three-and-a-half-inch second-hand telescope I'd bought for fifteen dollars. I used to sit in the back yard with Freeman Dyson and look at the planets and the stars. I would see my own moon of Jupiter. Dyson knew in far greater detail what was there."

Years later, Dyson gave a lecture titled "Mankind in the Universe" before the German and Austrian Physical Societies in Salzburg. He said, in part:

> The beginning of the space age can be dated rather precisely to June 5, 1927, when nine men meeting in a restaurant in Breslau founded the Verein für Raumschiffahrt. The V.F.R. existed for six years before Hitler put an end to it, and in those six years it carried through the basic engineering development of liquid-fuelled rockets, without any help from the government. This was the first romantic age in the history of space flight. The V.F.R. was an organization without any organization. It depended entirely upon the initiative and devotion of individual members. . . . In a strange way, these last desperate years of the Weimar Republic produced at the same time the splendid flowering of pure physics in Germany and the

legendary achievements of the V.F.R.—as if the young Germans of that time were driven to make their highest creative efforts by the economic and social disintegration which surrounded them. . . .

I now pass on to the year 1958. . . . I was one of a small group of scientists in America who were passionately interested in going into space but were repelled by the billion-dollar style of the big government organizations. We wanted to recapture the style and spirit of the V.F.R. And for a short time I believe we succeeded.

Our leader was a young physicist called Ted Taylor, who had spent his formative years at Los Alamos designing nuclear weapons. We started out with three basic beliefs. (1) The conventional von Braun approach to space travel using chemical rockets would soon run into a dead end, since manned flights going farther than the moon would become absurdly expensive. (2) The key to interplanetary flight must be to use nuclear fuel, which carries in each kilogram a million times as much energy as chemical fuel. (3) A small group of people with daring and imagination could design a nuclear spaceship which would be both cheaper and enormously more capable than the best chemical rocket. So we set to work in the spring of 1958 to create our own V.F.R. We called it Project Orion.

We intended to build a spaceship which would be simple, rugged, and capable of carrying large payloads cheaply around the solar system. We felt from the beginning that space travel must become cheap before it can have a liberating influence on human affairs. So long as it costs hundreds of mil-

> lions of dollars to send three men to the moon, space travel will be a luxury which only governments can afford....
>
> It is in the long run essential to the growth of any new and high civilization that small groups of men can escape from their neighbors and from their governments, to go and live as they please in the wilderness. A truly isolated, small, and creative society will never again be possible on this planet....
>
> We have for the first time imagined a way to use the huge stockpiles of our bombs for better purpose than for murdering people. My purpose, and my belief, is that the bombs which killed and maimed at Hiroshima and Nagasaki shall one day open the skies to man....
>
> It should be clear to everybody that Apollo is an international sporting event, in which science has only a subsidiary role. . . . Apollo will take men beautifully for short trips to the moon. But as soon as we are tired of this particular spectacle and wish to go farther than the moon, we shall find that we need ships of a different kind.

I once asked Dyson to describe Ted Taylor during the days of Orion, and he said, "I think of him mostly in La Jolla sitting under the stars and dreaming how we would go there. He loved the beauty of the stars. We sat there watching the planets go by. He didn't know much about them from a scientific point of view. He just loved to look at them. Did I intend to go, too? Oh, yes. Oh, very much so. Mars was 1965, if all had gone well, but I was more interested in Saturn, really. I said Saturn by 1970. One knows that Saturn has a lot of water available. You could refuel there. And the puzzles about Saturn are extremely interesting. For example, its satellite Iapetus is white on one side and black on the other. Why? That's what we'd like to know. Saturn would have been a two-and-a-half-year trip. Ted was fortunate in his marriage to Caro. She has great inner resources and would have made a perfect Penelope to his Odysseus. What was interesting was the amount of people and stuff we could have taken along in a payload of hundreds of tons. We could have stopped on the moon for a couple of months and really done some exploring. We always thought of the moon as being essentially a rather useless piece of real estate, though. Everything depended on whether we could find hydrogen there. It might have been a refuelling base. Ted, incidentally, was extremely good as a boss. He had time for everybody. He never got in a hurry. He had the gift of getting people to perform at their best."

Inevitably, a Super Orion occurred to Dyson—a ship immensely larger than the sixteen-story building under contemplation. So Taylor spent an afternoon figuring out how heavy Chicago was. What would it take to propel downtown Chicago through space at a few million miles an hour? Dyson had the answer. He imagined something a mile in diameter—using H-bomb propellants. The idea was to absorb all energy, then reradiate it at lower temperatures. Therefore, the pusher plate would be made of copper, which has high heat conductivity and would take the heat inside and then radiate it to space. This was a star ship, not a planet ship. Parts would be put into earth orbit by rocket, and the ship would be assembled up there. It would go out of the solar system and off to a nearby star in a voyage of a few hundred years. Its fuel would be a million hydrogen bombs. Taylor reports that Dyson would never smile when he talked about these things. "Dyson is a thousand years ahead of his time. He would disagree with that. He would say, 'Maybe a hundred.' "

Out to Point Loma, across the bay from downtown San Diego, Taylor and the others would take model Orions to test them in flight. Point Loma was a spectacular loaf of high cliffs and chaparral-covered hills reaching out into the Pacific. With shrieking seabirds overhead, small Orions would rise on clouds of flame and, generally, break to smithereens. The art became refined, though, and one day in 1959 a one-metre model blasted off and kept on going. Five movie cameras, operating at different speeds, followed the flight. The one-metre model—size of, say, a doghouse—had bombs inside that were made of high explosive. They were packed in cans strung together with Primacord, so the explosions could proceed from can to can. Each can had a metal plate on its bottom for "slight directivity" —to shape, somewhat, the charge. Boom. Rise. Boom. Rise. Boom. As the explosions came, they were orange and red and white and black, and they spread out hugely to all sides, with the small Orion sitting on them and climbing into the air. Boom. Orion went higher. Boom. Billows of smoke and fire. The thing was now a hundred feet in the air. Boom. If the ship happened to tilt on one blast, the next blast corrected the tilt. The flight was stable. Errors in timing or position were self-correcting. On up it went until the bombs were gone. A parachute opened, and the model slowly came down.

Orion attracted an impressive list of earnest supporters, among them Niels Bohr, Harold Urey, Curtis LeMay, Hans Bethe, Theodore von Karman, Arthur Kantrowitz, and Trevor Gardner. In 1961, Ted Taylor went alone to Huntsville, Alabama, to explain Orion to Wernher von Braun. Von Braun had no initial interest; he was just performing a courtesy. The Air Force had asked him to listen. While Ted talked with him—about temperatures, pressures, and other data—von Braun closed his eyes in apparent concentration, but Ted soon realized that von Braun was sound asleep. After a time, when von Braun's eyes opened, Ted turned on a movie projector and showed him, in slow motion as well as standard speed, flights of the one-metre model. Von Braun sat bolt upright. His face spread out in a big toothy grin. He asked for details. Could Ted give him certain data? Temperatures? Pressures? Von Braun was a vocal advocate of Orion thereafter.

Taylor worked on Orion seven years, the last of which were worrisome times, as the Air Force, which had taken charge of the project and had to present it as a military enterprise in order to get funds, moved to subvert Orion's purposes. "Whoever builds Orion will control the earth!" said General Thomas Power, of the Strategic Air Command. Power and a few others in the Air Force had in mind a space battleship with full-blown guidance systems and directional A-bomb explosives for bringing down missiles. It could run away from an enemy or it could turn around and take whatever might come—presenting its pusher plate to anything that came near it. Go ahead. Hit me. Orion could resist a megaton explosion five hundred feet away.

The limited-test-ban treaty of 1963 forbade nuclear explosions in space and in the atmosphere, and thus led to the indefinite suspension of Orion. "Just a few little twists of events and everything we were trying to do with Orion would have come through," Taylor has lamented. "It was, as Dyson put it, 'something more than looking through the keyhole of the universe.' It was opening the door wide."

TED TAYLOR believes in divine guidance. He believes there is an order in the world and some kind of benign influence assuring that everything will work toward some over-all good. "He always felt there must be good in what he was doing, and was able to think of some," Caro has said. "You would become pretty cynical if you were working on bombs and didn't believe in any sort of religious values." Ted recognizes God in whatever form. He is a Christian but does not believe that Christ was more significant than Buddha or any other manifestation of God. How does he draw the line between true and false messiahs? "It is like physics," he will answer. "You believe Newton because his work hangs together. You might not believe someone else." Sporadically, Ted goes to church. Of late he has been attracted to the teachings of Meher Baba—meditating, listening, or, as Caro says, "trying to plug into another level of existence: God, Krishna, Buddha, Christ, Muhammad, all one, proceeding in cycles." It was Clare, the oldest of his five children and mother of his grandchildren, Adi and Corwin, who led her father into knowledge of Baba. He is ready to go anywhere, though. Let all religions bloom. The dreams he has had since childhood of vast planetary discs rising proximate in his field of vision make him wonder if they are memories of things he has done, for he believes in an evolutionary spiritual process—appearance and reappearance, in different forms.

Ted is opposed to the making of rain. He does not want to destroy the unpredictability of the earth. His house in the hills is all-electric, but he did not plan or build it. He is an attractive man, tall, slim, almost wiry, with a head that is pointed not at the top but at the bottom, being broad in the forehead and sharp in the chin. A "Yield" sign. He wears dark suits, black shoes. He is neat—carefully arranges pencils and papers. Members of his family say he never gets dirty—not even on a camping trip. He is never late, anywhere. He may jump when a door bangs, but he is not a nervous man. His worry goes to his stomach. The aspect he presents is gentle and calm. The closest he will ever come to showing anger—by the testimony of all who have lived with him—is to hold his breath and shut his eyes. One of his daughters has said, "I was glad when he spanked us, because it never hurt." His hair is dark and has flecks of gray in it now. His eyes—the most notable of his features—are brown, rich in texture, full of youth and pure inquiry. When they focus on someone he is talking to, they deliver an assurance of authentic interest, and when his thoughts leave his immediate world his eyes seem to see nothing and to suggest the worlds they cover. "He has done some alarming things in traffic," his wife has said. "He does not notice where he is going. If an exhaust pipe fell off and was dragging on the road, he would not notice. He would not notice a flat tire. If you concentrate as hard as he does, you have to lose something. He doesn't know what is going on, but he thinks like the dickens. He has no mechanical aptitude."

"What would you do if the car broke down?" I once asked her.

She said, "We'd go and find a strong man to push it. Some people are mechanical. Ted is theoretical."

He drives a Volkswagen, and I got into it with him one day to go to his favorite Chinese restaurant, which is on Connecticut Avenue in Washington, near the Maryland line. After we went by the restaurant the first time, I said to him, "What was the name of that place again?"

He said, "Peking."

I said, "It was back there—two blocks."

He turned around, started back, picked up speed, shot past Peking and on toward the center of the city. (He explained later that he had been thinking of ways to separate uranium and plutonium oxides.) On the third pass, he circled the block until he found a place where there was enough open curb space for an Allied moving van. He worked his little bug into the center of the space, slowly. He cannot park a car. He has trouble starting cars, too. I have seen him try, and try again, to start one with a hotel key.

While making preparations for one of his numerous journeys in connection with his work on nuclear-materials safeguards, Taylor asked Carl Goldstein, of the Atomic Industrial Forum, if he could arrange a visit to a certain fuel-fabrication plant in Pennsylvania. Goldstein called the company, which said it would check Taylor out and call back. Several days later, Goldstein got this message: "We've looked into Taylor. The man is a genius. He can't come here."

Physicists seem to talk about each other as much as actors do. Those physicists who have known Taylor are not reluctant to talk about him.

"Like many of us, he was at first fascinated with the technical problems, and only later it caught up with him. I went through the same experience. What catches up with you is that you find out the people who are in charge of these things aren't as wise as they should be."

"Some people think that one man or two could not make a bomb, but Ted Taylor manifestly could do it if he wanted to."

"He is worried about his children and the world he would leave them in. The fact that he developed bombs has given him a burning desire to mitigate their effects."

"The difference between Ted and someone like, say, Manson Benedict, at M.I.T., may be a matter of personality. Ted worries about homemade bombs and Benedict does not. There are people who have an urge to look for troubles and there are people who find it uncomfortable to do so."

"His credentials are such that we know he is not an irresponsible crank."

"He knows intuitively the forces on other people, knows their weaknesses."

"He is not gullible. He is not naïve. He is willing to be naïve and gullible on things that do not matter."

"He is willing to accept a simplicity of approach where things may be more complex."

"He is just a little bit wild."

"Ted is a little cracked—but just a little."

"Like the Liberty Bell?"

"I don't understand your allusion, but if I did I would probably agree."

ONE afternoon at the cabin in Maryland, Taylor moved his slide rule for some minutes and began to mutter to himself. "Let's see. It's easier to accelerate something short than something long, so let's try out a five-and-a-half-inch projectile. Pi *r* squared times the length—five and a half inches—times two point five four times nineteen is about ten to thc fourth. Ten to the fourth divided by sixty times thirteen point three is twelve and a half. The square root of twelve and a half is three and a half. Yeah! That's all right! We've come up with a nifty design here."

He was designing, freehand, a gun-type fission bomb. It was a thing about three and a half feet long, and it had seven principal components: projectile, target, initiator, reflector, propellant, container, firing system. It was a crude bomb. It weighed five hundred pounds. The materials and knowledge necessary for its making were all available in public markets and public print, with the exception of the fissile material, uranium-235, which was now becoming available in, among other places, the nuclear-power fuel cycle. Time and again, while he was sketching out the design, he would stop and say something like "There's a level of detail here to which I can't go with you and which would make the whole thing much easier, but this way will do."

The basic requirement was that two subcritical

pieces of metallic uranium come together very quickly. In the presence of a surrounding reflector, which could be made of steel or any other material that would tend to reflect neutrons back into the core, the assembled uranium would be supercritical. The nearly instant result of a fission chain reaction, begun after the assembly became supercritical, would be a nuclear explosion.

Design began with the size and shape of the two pieces of uranium and their relationship to the intended reflector. To find out how much steel to use around how much uranium, one would look at the Los Alamos critical-mass summaries. For example, the critical mass of U-235 inside three inches of steel was twenty-six kilograms; one inch of steel, thirty kilograms; a half inch of steel, forty kilograms. The figures referred to spheres of metallic uranium but would be much the same if the metal were formed into a squat cylinder. A designer would want at least ten per cent more than a critical mass, because after that efficiency rises dramatically, with the promise of higher and higher yields. Taylor decided on a forty-kilogram cylinder of uranium, which would be five and a half inches in diameter and five and a half inches long. The cylinder consisted of two parts, one—a sort of plug—fitting inside the other. The dimensions of the plug were two and three-quarters inches in diameter and five and a half inches long. At detonation time, the plug—the projectile—would be fired down a gun barrel and into the larger piece, the target. This assembly, surrounded by a three-inch steel reflector, would be sufficiently supercritical.

While the plug was at the far end of the gun barrel, the whole arrangement would be subcritical. Such, in general, had been the design of the Hiroshima bomb.

Questions came to mind. How would he mold the two pieces of uranium?

In magnesium-oxide crucibles, he said—which, in turn, had been shaped by hand.

There would have to be a hole in the reflector wall, so the uranium projectile could go through the reflector and into the uranium target. Why, after the projectile had been successfully fired into place, would neutrons not pour out that hole, fizzling the explosion?

Because the projectile would have a piece of steel screwed to it, riding piggyback behind it, and this would fill in the reflector wall.

Where would someone get an appropriate gun barrel?

Buy an old cut-up three-inch naval gun barrel from a military salvage yard, he suggested. You could use a surplus bazooka, for that matter. The reflector could be made from segments of larger guns, or from a cast-iron block with a hole drilled into it—or from tin. Tin melts at two hundred and thirty-two degrees centigrade and could be molded at home. Solder, which is tin and lead, might be a shrewder choice. The purchase of so much tin might raise an eyebrow, but no one would think twice about solder.

Water could be used as a reflector, reducing considerably the weight of the bomb. The gun device alone could be fired in a swimming pool, a water tower, even a toilet. The water would reflect enough neutrons for supercriticality. A nuclear explosion would result. So far as Taylor knew, such a method had never been described or tested in any weapons program.

Plutonium is not right for a gun-type bomb. The isotope plutonium-240 cannot be completely separated in any practical way from plutonium-239, and plutonium-240 fissions spontaneously. It does so with such vigor that if a plutonium projectile were fired toward a plutonium target, neutrons, which travel at speeds in excess of fifteen million miles an hour, would jump from the one to the other before a proper assembly could be achieved. The result would be a fizzle yield. This problem—predetonation—is serious enough with uranium, because of the unavoidable presence of cosmic rays and other stray neutrons. That is why the assembly of the separate parts has to be fast. The projectile travels at a rate of about five hundred feet per second.

I asked Taylor how he would accomplish that in the bomb he had been fabricating in his mind.

Plain gunpowder, he said. The projectile's weight was more or less thirty-five pounds. Less than two ounces of gunpowder could make it move five hundred feet per second down the gun barrel. Put the gunpowder into a small plastic bag. Run a wire from a flashlight battery to the bag. Connect the battery to an alarm clock. In place of the clock, you could use a preset mechanical fuse, which can weigh as little as a fraction of an ounce. Hundreds of types are commercially available. A barometric fuse would do. So would a radio signal.

When the uranium projectile and the uranium target conjoined, stray neutrons might be on hand to initiate the explosive chain reaction, but no one would depend on that. Instead, the designer would incorporate an initiator—an instant source of neutrons. There are many ways to make initiators, all involving two or more materials that, brought together, will emit neutrons. The initiator must emit nothing until the assembly has become supercritical. One way to do this would be to buy some lithium from a chemical-supply house, shave off a little with a knife, and glue a lithium wafer to the front of the projectile. A tenth of a curie of polonium-210, a substance in common use in university physics laboratories, can be glued to the steel at the far end of the target. When the lithium hits the polonium, a burst of neutrons will scatter in all directions into the uranium.

The container of the bomb need not be a torpedolike steel jacket with fins. For the bomb Taylor conceived there at the cabin, a garbage can would do, a clothes hamper, a golf bag. Its over-all size was three feet six inches by eleven inches, which was its maximum diameter at the exploding end.

The Atomic Energy Commission's *The Effects of Nuclear Weapons* and Samuel Glasstone's *Sourcebook on Atomic Energy* would be helpful general texts for anyone attempting to make such a device. Taylor guessed that one person would need a few weeks to complete the project. Any college textbook on the theory of fast breeder reactors would be helpful, because bomb theory and fast-breeder-reactor theory are much the same. To predict yield, one could turn to *The Los Alamos Primer.*

I asked Taylor what sort of person could read and understand *The Los Alamos Primer.*

He said anyone who had got a fairly good grade in an introductory course in reactor engineering or reactor theory, even at the undergraduate level.

I asked him what might be the yield of the bomb he had just conceived.

He said a kiloton. He had deliberately thought up a highly inefficient bomb—but one that was vastly lighter and more compact than the bomb that destroyed Hiroshima. He said there were ways to reduce the weight even further, but at some expense in yield. An aluminum gun barrel and a block-aluminum reflector could be used, for example. Just order a pipe and a block from Alcoa. The uranium core would have to be increased some twenty per cent, but the total weight of the bomb would come down to two hundred pounds. The yield would drop from a kiloton to about half a kiloton, he guessed. A tenth of a kiloton, however, would be enough to bring down the World Trade Center.

IN a Hertz car, Taylor and I, unannounced, once drove in the lonely road that leads to the New York State Atomic and Space Development Authority's storage facility, near West Valley, New York. We turned off the engine and sat and talked, absorbed in the plain fact that the capacity of this relatively small and isolated building was two thousand kilograms of plutonium—core material sufficient for hundreds of implosion bombs. A car was parked near the gate in the chain-link fence. No activity of any kind was discernible. Taylor lit a cigarette. He said, "Is it worse, or better, if it happens sooner, or later? Someone, somewhere, stealing material and making a bomb. I think it is better sooner than later. I would not be surprised if it happened tomorrow, and I would not be surprised if it didn't. I would rather have it happen tomorrow than ten years from now, when so much more material will be floating around. I've sometimes found myself actually wishing it would happen—perhaps an unusual group of people, ten years ahead of their time, who would sort of jump the gun on everybody else before a large number of organizations could be ready and poised to do the same thing. If it happened just once, the lid would be slammed shut. As shut as possible, anyway. In another ten or fifteen years, it will be too late. If I were to read it in the papers tomorrow, I'd be frightened and alarmed, but—if it aborted and very few people got hurt—I'd be, in a sense, happy."

He went on to say that by no means was he an unqualified pessimist about safeguards. He believed they could be developed—not only in the United States but also around the world—in a way adequate to control the inherent peril of special nuclear materials. He did say he thought it was already too late to prevent the making of a few bombs, here and there, now and then. Society would just have to take that, and go on. None of this was said with the least trace of cynicism or despair—two characteristics that do not seem to be in his repertory. Seeking what a lawyer friend of his called "a mechanism to make a difference," Taylor founded in 1967 a firm called International Research & Technology, with a charter purpose to serve as a private monitor of nuclear-materials safeguards. Over the years, it grew until it had some twenty-five physicists, mathematicians, engineers, political scientists, sociologists, economists, and lawyers, but at first it was just Ted and Caro. For two years, they lived in Vienna, where he observed the workings of the International Atomic Energy Agency. They were supported by consulting contracts with the A.E.C., the Stanford Research Institute, and General Atomic. If Taylor is something of an alarmist, he is also an instinctive believer in the good will of man, and he came away with a conviction that international safeguards—inspectors based in Vienna and travelling the world to see that nations and individuals do not misappropriate materials—could be made to work successfully. He was impressed, for a beginning, with the way Russians, Argentines, Yugoslavs, Englishmen, and Americans functioned cohesively in the international agency. He did not feel a sense of hopelessness in the fact that roughly half of the weapons-grade nuclear material in civilian power fuel cycles is outside the United States. His sense of dismay began to increase rapidly after his return, as he observed the rate of growth of the nuclear industry and the comparative inattention paid to safeguards, so he became a more or less permanent safeguards critic, although the effort would always seem frustratingly uncreative to him, a somewhat tedious and burdensome duty. "Everything is a matter of probabilities," he said now, in the Hertz car. "This is true in all physics. It is the core of the safeguards problem." He looked over at the small metal warehouse, a car parked beside it. He looked from the seven-foot fence, topped with barbed wire, to the broadly cleared perimeter and the surrounding woods. He could almost see a chopper landing, or a truck full of masked commandos drawing up to the fence. "What is the measure of likelihood that people will try to get into this storage facility, take plutonium, and make some bombs?" he said. "You know it's not one in one. And you know it's not ten to the minus·ten."

The Atomic and Space Development Authority's storage facility, being the home of the largest private stockpile of plutonium in the world, should be an uncompromising example of the state of the art of safeguarding nuclear materials, and that is exactly what James Cline thinks it is. Cline, chairman of ASDA, worked on the design of the place, and his attitude toward plutonium thieves is "Let them come—this is one place where they are going to get programmed into something they'll remember." The warehouse is rigged with, among other things, the electronic systems that are used to protect armed missiles. Any surface—roof, floor, walls—breached by an invader will set off a silent alarm. Even if an invader could somehow get inside the building without touching anything, an alarm would react. To achieve such "ultrasensitivity," as Cline calls it, many kinds of systems are used, some electrical, some electronic. ASDA has bought everything available.

Cline's office, three hundred miles southeast of the storage facility, is on Forty-fifth Street, in Manhattan, where I called on him one day and asked him why the warehouse was so isolated, why almost no one seemed to be there, and what would happen if the place was attacked.

There was nothing defensive about his response. A tall, amiable man—a nuclear and electrical engineer who once was on the A.E.C.'s reactor-development staff—he seemed somewhat nervous, but not about his warehouse. He said that people coming and going were exactly what a well-safeguarded storage facility would want to avoid. People panic. Electronic systems do not panic. You can't hold a gun to the forehead of an electronic system. When people are around, certain alarms have to be turned off. The Brink's robbers studied Brink's, then decided to hit the place at 7 P.M., when a minimum number of employees were on duty but the vault was still open and the alarm was shut off.

Most days, the only person at the ASDA storage facility was Joe Merkley. It was his car you might see parked outside. Merkley was the superintendent. He maintained instruments and kept the building warm. No, he was not a janitor. He had two years of engineering and two years of business administration. When Merkley arrived for work in the morning, he could not even let himself in. He was electronically coupled to other people, miles away. Systems in the facility were zoned. When Merkley was inside, only the zone he was in was shut down. The place was even safer when he was not there and the alarm systems had it all to themselves, which was three-quarters of the time. The fence was for completeness, that's all. Go ahead and pole-vault it. The response, to any kind of invasion, would be fast. How fast? Damned fast. Before anyone could even orient himself to the interior, groups would come. Who? Every enforcement resource you could possibly bring to bear. Distances were short and people were prepared. The alarms were silent, and the invader would not know they had gone off. The response would be, for all practical purposes, immediate. Anyone who wanted plutonium should look elsewhere in the fuel cycle, not at the ASDA storage facility.

The principal room inside the facility was laid out for a forest of about a thousand drums, Cline said, each containing a ten-litre flask of plutonium-nitrate solution, which, in turn, contained about two kilograms of plutonium. The drums were steel and nearly six feet tall and lined with thick concrete—top, bottom, and sides. The lids displayed tamperproof seals in different colors—ASDA seals, owners' seals, and International Atomic Energy Agency seals if an I.A.E.A. inspector had been there when the drum was assayed and shut. Once sealed, the drums were like the bricks at Fort Knox, Cline said—objects of known content. A single drum weighed more than a ton, and no one was going to go anywhere with it. The concrete plug in the top alone weighed three hundred and fifty pounds. Besides, what would a bomber do with the nitrate if he got it? He would need "all the wherewithal of a small Manhattan Project." Cline was not worried about theft in the United States, but he hoped the facility had set an example for other organizations to follow in working with the I.A.E.A. He was interested in strong international safeguards, and did not think cooperation with the I.A.E.A. had developed as well as it might. As it happened, the plutonium at the ASDA facility was well below capacity, because quantities of it had recently been sold to England, Germany, and Italy. There would in a few years, though, be so much plutonium around that it would inevitably become a commodity. There would no doubt be a plutonium exchange—plutonium futures. There would soon be more plutonium in private storage than in all the bombs of NATO. But anyone, then or now, who so much as approached the ASDA storage facility—even anybody found on that road outside it—would have a good long explanation to create.

Taylor started up the Hertz car and drove slowly away from the facility. We had been sitting there about half an hour. Nobody had asked for an explanation.

Small flatbed trucks had taken the plutonium from the Nuclear Fuel Services reprocessing plant to the ASDA storage facility, following a route of about four miles—going through woods and beside fields, a left, a right, another left, past a mobile home, a farm that sells maple syrup. The trucks carried ten birdcages at a time. Each birdcage weighed about four hundred pounds, and held a ten-litre flask of plutonium nitrate. A hijacker taking such a truck would have enough plutonium for four or five bombs.

Larger trucks, owned by big trucking companies, take plutonium and fully enriched uranium on journeys of a thousand miles and more. These major shippers—Tri-State, McCormack, Baggett, Pacific Intermountain Express—are the first to say that in the matter of safeguarding fissile nuclear material transportation is the most vulnerable part of the fuel cycle. Everybody—from nuclear academics to utility people to the A.E.C.—agrees that this is so. William Brobst, the chief of an A.E.C. department that deals exclusively with transportation, says that the situation has to be considered in the wider context of the trucking industry as a whole.

"In trucking, there is constant pilferage, day-to-day theft, 'shortages,'" Brobst said one day over lunch in the A.E.C. cafeteria in Maryland. "Of all the freight of any kind that is shipped in this country, one to two per cent disappears. Holy mackerel! Is this going to happen to our nuclear materials? The trucking business is a mass of crime. Eighty-five per cent of its thefts are by authorized people—that is, by people who belong there. Four out of five truck hijackings involve collusion—the drivers are in on it. That's the situation. The nuclear business has vaults, fences, alarms. They have great machines to count atoms. Then they take all these atoms, lump them into a drum, and toss them out into crime.

"The earliest move toward civilian transportation safeguards was in 1970, when the A.E.C. woke up to the magnitude of the problem and began requiring signed receipts for shipped material. By now we're living in a terrorist society, O.K.? That may be an exaggeration—but you respond, anyway, to threats, and the terrorist thing has grown almost exponentially. It is not just *making* a bomb that worries us. We don't want even the *threat* of a bomb. Still, I would like to see the transportation problem solved within the context of our national transportation system. The question is how to ship the stuff and still use the civilian transportation industry. The law prohibits the government from competing with private industry. The law could change. But I'm not for that. If you're willing to pay for it, you can get a lot of added protection—for example, exclusive-use vehicles instead of ones that carry all sorts of other freight and nuclear materials, too. You can have as much safeguards as you're willing to pay for."

Brobst had a sandy, somewhat pointed beard, which seemed to make his points with him as he went along. He said that one way to reduce collusion between drivers and hijackers was to insist on two drivers. He said that radio transponders could be set up on a truck in such a way that if a driver did not regularly press a button to show that all was well an alarm would go out to police. He said that a truck could even be equipped with "an automatic poison-gas system—if someone turns the wrong dials, the truck sets off three skyrockets and shoots chlorine into your face." A hijacked truck could destroy itself. It could broadcast an automatic SOS, blow off its own wheels, crumple its axles, and sink down on the road like a sick elephant.

Early in 1973, the A.E.C. ruled that weapons-grade material above a small number of grams should no longer travel in passenger aircraft. At the same time, the Commission presented for comment a set of proposed new regulations, reflecting, if nothing else, the growing need for materials safeguards. Under the new regulations, published in their effective form at the end of the year, inventories of weapons-grade material, wherever it happens to be, have to be made every sixty days for uranium-235 and plutonium-239. Depending on the type of material, the amount of MUF (material unaccounted for) can vary, but in no case can the MUF exceed one per cent. The specific areas within plants where weapons-grade material is stored or processed must be protected by special alarms, barriers, guards.

All companies that handle more than two kilograms of plutonium-239 or five kilograms of uranium-235 must prepare, for A.E.C. approval, a comprehensive physical-security plan designed to discourage industrial sabotage and to protect the materials from theft. These plans—involving a great number of things, from fencing systems to frisking procedures to communications with local police—are intended, collectively, to be the basis (as far as physical protection is concerned) of the national safeguards system. Vehicles making any kind of delivery in protected areas are to be escorted and their drivers must be searched, but the vehicles themselves are not subject to search.

Because transfer points are where most trucking pilferage occurs, trucks carrying fissile material over long distances must go straight from origin to destination, stopping only for food or rest or maintenance. They must have two people in the cab. Truckers must either use a "specially designed vehicle which reduces the vulnerability to diversion" or send along two armed guards in a separate vehicle capable of instant radio communication with the truck. (Brobst had been against arming truck drivers themselves. "That could be dangerous," he said. "It's like arming your wife. Furthermore, most truck drivers are probably armed already.") Top and sides, trucks must be marked. Containers must weigh more than five hundred pounds if they are carried on open trucks or in railroad cars or on ships, but portable containers may be used in trucks that are locked. Cargo aircraft can continue to carry fissile material in portable containers, but the containers must be sealed and locked. Trucks must report their position by radiotelephone every two hours or, if they are out of transmitting range, by conventional telephone at least every five hours. All measures taken together are intended to protect materials in transit from any attempted theft "short of a significant armed attack."

"It all sounds very impressive," Ted Taylor has said. "But look again at that phrase 'short of a significant armed attack.' Are transportation safeguards supposed to deal only with *insignificant* armed attacks? Does the same ground rule apply to sites where enough plutonium for a few hundred bombs is stored? Where, moreover, are the specific standards that one would expect a regulatory agency to use in deciding whether to approve a licensee's physical-security plan? What is the design of a 'specially designed truck'? How long should the design prevent a group of hijackers from getting inside? How long should it take to bring law-enforcement agents to the scene of a theft? How big a force? What weapons should plant guards carry? How many guards? What are their orders? To delay? To kill? To capture? If delivery vehicles going into protected areas are not to be searched, why could such a vehicle not be used as a Trojan horse? In hearings to justify its budget for fiscal 1974, the A.E.C. said, 'Almost no standards exist in the materials-protection area, and in many cases the basic data needed to develop such standards have not been developed.' I couldn't agree more. *Money* is better protected than uranium or plutonium. I can find no evidence that the type of safeguards system the A.E.C. is now calling for could deal effectively with nuclear thefts by professional criminals with talents and resources similar to those that have been successfully used for major thefts of other valuables in the past. Until such a system is evidently in operation, we are all being asked to take what I consider to be an unacceptable and, I should say, unnecessary risk. I have by now lost all patience with people who say that criminals who really want the stuff are going to be able to get it no matter what the A.E.C. and the industry do. As far as I can tell, the A.E.C. and the industry have just begun to try to do anything about this. I doubt if people in the nuclear-power industry expect ever to spend as much as one per cent of their resources on measures designed to prevent nuclear theft."

Perhaps by coincidence, at about the time when the A.E.C. published in effective form the regulations that had been introduced nine months earlier, the Comptroller General of the United States sent a fifty-two-page "Report to the Congress" on "Improvements Needed in the Program for the Protection of Special Nuclear Material." The General Accounting Office, acting within its responsibility to assess performances by other government agencies, had checked on the protection of weapons-grade uranium and plutonium in three civilian facilities. No names were named. The plants were called Licensees A, B, and C. Each contained a great number of kilograms of S.N.M.—special nuclear material. What the G.A.O. agents found, at two of the plants in particular, were "weak physical security barriers, ineffective guard patrols, ineffective alarm systems, lack of automatic-detection devices, and lack of an action plan in the event of a diversion of material." The report was enough to dismay the A.E.C., let alone any private citizen who might have imagined that private companies with bomb-grade material would follow to the letter any A.E.C. regulation that happened to exist, for even the comparatively mild procedures that have been required had been frequently and flagrantly ignored. The report mentions that the A.E.C., reviewing the General Accounting Office's recommendations for improvement, "said that it has taken, or is taking, actions to implement them." These are examples from the report:

> We believe that Licensee A's security system was significantly limited in its capability to prevent, detect, and immediately respond to possible S.N.M. diversions or diversion attempts. The security personnel consisted of a part-time security officer and four guards. . . . Each guard on duty carried a .38-caliber revolver. The weapons qualification scores, however, showed that none of the guards had met A.E.C.'s requirements. . . . The watchclock tapes which recorded the guard patrols indicated that the guard on duty did not vary the time or route of his patrol. When he was not making watchclock checks, the guard was located in a small guard post. . . . The guard could not observe

about 80 per cent of the general plant area from his post. . . . The fence could be easily disassembled because the nuts, bolts, gate hinge pins, and wire used to fasten the fence mesh to the fenceposts were not secured by welding or peening. . . . One of the storage areas—which housed both S.N.M. scrap and S.N.M. of high strategic importance—was a prefabricated steel structure. . . . We tested the impediment value of the panels with an adjustable-jawed wrench. Within 1 minute we were able to remove five metal screws from one of the panels. At this point the only impediment to entry was a small rivet which, in our opinion, could have been forced manually. . . . We also tested the impediment value of the sheet-steel panels by cutting a sample of the panel. Within 30 seconds we were able to make a 19-inch cut with an ordinary pair of tin cutters. . . . The garage-type door could be opened with little effort because its lock had been broken and the door could be opened without activating the alarm. . . . The other storage area, which provided limited protection, was smaller and constructed of cinder block. . . . On the side of the building facing the perimeter fence were two unalarmed vents leading inside. This side of the building, which was about 16 feet from the perimeter fence, was not visible from the guard post and, according to a licensee official, was inspected only once a month. One of the vents was located about 2 feet from the ground, measured about 18 by 30 inches, and was secured on the outside by louvers and an ordinary window screen. . . . The licensee's manager for safeguards and accountability concurred that with little effort the louvers could be pulled out by hand and that the inside screen could be manually forced, providing access to the building interior. The other vent was . . . secured on the outside by louvers and on the inside by a piece of thin sheet steel fastened to the cinder blocks by four bolts. . . . Again the licensee's safeguards and accountability manager concurred that this vent could easily be pulled out without tools and that the sheet steel could be forced manually. Portable S.N.M. was readily accessible within the cinder block warehouse. . . .

The integrity of the front wall [at Licensee B's plant] was impaired in that none of the windows were laminated, sealed, locked, or alarmed; windows (frames and glass) were nonexistent at two openings; one of the doors was open with a broken seal attached; and none of the doors were alarmed. The rear of the building was windowless, did not have protective fencing, was not visible from the guard station, and was not routinely patrolled by the guard. During our tour around the building, we observed a screen covered with plasterboard which was used to secure an opening. . . . The screen was held in place by three toggle bolts. Within 15 seconds and using no tools, one person was able to remove the bottom toggle and open the screen to about a 45° angle. . . . The opening led directly into an S.N.M. storage room which was locked but not alarmed and which contained significant quantities of S.N.M. stored in easily portable half-gallon containers. The opening was cemented and sealed within 1 hour after our tour. . . . We tested the accounting controls by comparing seal numbers provided by the guard lieutenant with those on the doors and gates and found that only 5 of the 10 were correct. . . .

Licensee C had established liaison with local law enforcement authorities and had developed an informal plan intended to provide assistance in the event of an emergency. The licensee's arrangements called for hourly communication checks to the local police. If the police failed to receive the call at the designated time, they were to contact the licensee by radio or telephone and, if contact could not be made, were to respond by dispatching a squad car. In a test of the effectiveness of this arrangement, we found that the local police attempted to call the licensee within 15 minutes after the licensee failed to call at the appointed time; the squad car which was dispatched, however, went to the wrong facility 14 miles away.

The Comptroller General's report reproduced the ratings given these plants by A.E.C. inspectors, who had found them uniformly "good." Protection of openings: "good." Emergency plans: "good." Security of material in storage: "good."

Ted Taylor's firm, International Research & Technology, now a subsidiary of the General Research Corporation, has branched wide from its original and continuing preoccupation with safeguards—going into urban-transportation planning, solid-waste management, the economics and technology of air- and water-pollution control, and a technological assessment of the functioning of the United States Postal Service. I.R. & T. set the specifications for three steam-engine buses that operated experimentally in Oakland, California. In a study for the Department of Justice, I.R. & T. discovered certain patterns in bank records of the payroll checks of a company in Detroit, bringing into the open four extremely surprised usurers. Taylor wishes he could do more of the creative things and less nuclear watchdogging, and such dreaming has led him to what could be the ultimate safeguard. A source of vast energy alternative to nuclear energy would clearly eliminate the safeguards problem, because the nuclear industry would disappear. Making none of the evident stops, he has all but ignored the winds, the tides, and falling water, the molten core of the earth—interesting but insufficient alternatives. What he would like to do is build greenhouses a hundred miles in diameter. Looking over all energy sources some time ago, he was surprised to find that plant fuel—wood, alcohol, charcoal—was seldom even listed. This seemed odd, since plant fuel was the principal source of energy in the world before 1850. He calculated the energy that could be got by burning all the wild plants now on earth, and he found it would amount to twenty times the present total energy consumption. Cultivated crops would greatly increase that yield. An acre of corn was equal to about five tons of medium-grade coal. An acre of sugarcane was four times as good as that. There would be two big problems, though—land area and agricultural pollution. A hundred million acres under cultivation would be needed to supply the present energy needs of the United States, and that was five per cent of the land area. Moreover, fertilizers would be leaching into the streams. Therefore, the entire process should be contained in a greenhouse, or greenhouses, which would reduce the required acreage to twenty million and keep the fertilizers out of the streams. Glass was too expensive. Polyethylene would do. It was a simple organic compound, which, when ultimately replaced and incinerated, would turn into water and carbon dioxide. Greenhouse structures would be inflated and self-supporting, over frameworks of aluminum or light steel. Nevada would be a good site, because of its almost uninterrupted sunlight. The great greenhouse could be built in the valley where—in Air Force terms—the giant mushrooms once grew. Water would not be a problem, because it would be recirculated, and, moreover, only a fiftieth as much would be required as for normal agriculture. Burning chopped cane would supply heat to steam generators. Trapped filtered smoke and ashes would yield potassium, phosphorus, and nitrogen as fertilizer. Carbon-dioxide exhaust from the burning cane would go back in pipes to the growing cane, where, with sunlight and water, it would form starch and sugar. It would be a closed system—sunlight coming in, electricity and heat going out, heat for houses and for industrial processes. That the heat would thus be released to the environment would make no ecological difference, because the system would be converting sunlight, using heat that was there anyway. The polyethylene would be five-thousandths of an inch thick, enough to stop a hundred-and-twenty-mile wind, enough to stop hail. There would be two walls, six inches apart, for insulation. This great earth tent of plastic would cost sixty-six hundred dollars an acre and require replacement every five years. For enough greenhouses to supply all the energy needs of the United States, the metallic infrastructure would cost two hundred billion, and the plastic bill would work out to another fifty billion a year. Cheap, according to Taylor. "The development work would take much less money than for other systems. You don't have to build a reactor. You don't need nuclear physicists." See *The Restoration of the Earth,* by Theodore B. Taylor and Charles C. Humpstone, Harper & Row, 1973.

I DON'T think Taylor heard or noticed the rain, a bass-drumming rain, on the roof of the cabin. He was audible enough above it, though, and he was imagining someone with a glove box, a ceramic crucible, and a hundred-dollar electric furnace molding a hemisphere of plutonium—beginning the construction of an implosion system. The metal cools, and another, identical hemisphere is made. The two together form a sphere about the size of a grapefruit. The dead center of an implosion system is known as the pressure spike, and it is ordinarily filled with an initiator. In a bomb made with plutonium from a power reactor—plutonium stolen from the nuclear-power fuel cycle—no initiator would be needed, because enough plutonium-240, which fissions spontaneously, would be present to do the job. Government bombs contain very little of it, because its spontaneous fissioning would set off a chain reaction too early in the implosion and thus lower considerably the yield of the fireball. The amount of 240 that exists in a given amount of plutonium is determined by the length of time the fuel elements have remained in a reactor. Hence, the fuel elements in the government's plutonium-production reactors are removed for chemical separation before much 240 has accumulated. Civilian power reactors allow their fuel to burn a lot longer—so the resultant plutonium includes an amount of 240 very likely to set off a bomb too soon after the moment of criticality. This may lower the yield—but, even so, not to a level unacceptable to a clandestine bomber, who receives as a kind of dividend the presence of an automatic initiator.

Now the reflector. What is needed is a good neutron reflector, and to learn what is a good neutron reflector you can look, for example, in Glasstone's *Sourcebook on Atomic Energy,* under "Nuclear Reactors: Reactor Moderators and Reflectors." You could use natural uranium, steel, copper, magnesium, lead, aluminum, beryllium, water, solder, or wax. Two stainless-steel mixing bowls could be lined with wax and soldered together around the plutonium sphere. A three-inch thickness of wax will reflect as many neutrons as an inch and a half of steel, but you will pay a price in yield, because the explosion knocks the wax apart faster and the net explosive force declines. You could buy a pressure vessel of the sort that is used to lower instruments into the deep sea. It has steel walls two inches thick. Open it and put the plutonium sphere inside. Alternatively, make the reflector out of beryllium—the material that Taylor used for the reflector of Scorpion. Beryllium is among the most poisonous nonradioactive inorganic materials on earth. One of the lightest elements, it is less dense than aluminum, but it has so many atoms per cubic centimetre that it makes an especially good reflector. Beryllium, in fact, has more atoms per cubic centimetre than any other element. It is, as well, a good neutron scatterer. The critical mass of plutonium is smaller in a beryllium reflector than in any other reflector of comparable thickness. Beryllium costs about a hundred dollars a pound from, for example, the beryllium division of Kawecki Berylco Industries, in Reading, Pennsylvania. The metal is brittle and hard to work. But if someone wants a bomb that he can easily pick up and carry, he may want to go to the trouble of making the reflector from beryllium.

The over-all minimum diameter for an implosion bomb is secret. The plutonium core can be as small as a billiard ball, and the beryllium reflector around it can be less than an inch thick—those figures are not classified. The art of implosion design lies mainly in the high explosive that surrounds the reflector and the core, and it seems ironic that among fission-bomb secrets those which have inspired the intensest activities in international espionage have had to do not with special nuclear materials but with ordinary TNT—its amounts and its design arrangements. "What the lay bombmaker would do here is tough to predict," Taylor said. "He would know, probably, that if he added, say, two feet of high explosive he would get enough compression for a good yield, but he would have a bomb that was roughly five feet in diameter and weighed several tons, like the one that was dropped over Nagasaki." In a general way, some government reports offer insight, he continued. From articles in technical journals on the design of shaped charges, one could figure out how much implosive energy is necessary to increase the density of plutonium to the critical level. Then, in textbooks on high-explosive technique, one could find out how much energy is in a pound of high explosive. The Nagasaki bomb was completely covered with TNT in large blocks that were cemented on, and the gaps between them were filled with wadding paper. Today, one would use a modern plastic explosive, such as C4, which consists of TNT plus a plasticizer. It has the consistency of putty. It can be formed into anything. It is safe to handle. To buy it, go to the Hercules company and pose as a miner or a professor. Plain dynamite or TNT would be all right but unreliable.

Around the reflector, the high explosive is kneaded and formed, by hand. It is hard work. The stuff has the same general feel as putty. The idea is to achieve a uniform thickness. The bombmaker works at first by eye, but when he gets near the thickness he wants he checks it with a measured wire, poking the wire into the high explosive until it hits the reflector. A Geiger counter sits close by, and if there has been a miscalculation the clicks will reveal it. The plutonium core, inside its reflector, must, of course, be subcritical, but only minimally. High explosive is a fairly good neutron reflector, so, as the C4 is packed on, the bomb as a whole will move closer to criticality. If the Geiger counter makes it evident that criticality is too near, the whole structure has to be disassembled, recalculated, and rebuilt. Given the remote chance that the C4 might explode accidentally, a group making a bomb would do well to send one of their number off into another room to mold the high explosive in shells on an upturned salad bowl. The cores and high-explosive components of government-owned bombs are always made separately. A lone operator would hardly bother, though, for an accident with the high explosive would kill him and his plans.

"To detonate the high explosive, the bombmaker has a very wide variety of options," Taylor said. "What he chooses depends on how much he knows about high-explosive technology and how much he cares that the bomb will perform in the best possible way. High-explosive lenses are commercially available in a wide variety of types. Accurate explosive fuses can also be bought commercially. If you don't care whether you get a tenth of a kiloton one time and five kilotons another time, you can be much less fussy about the way the high explosive is detonated. This is a very sensitive subject, and that is all I can say."

After detonation, about a third of the explosive force of the C4 goes into a compressive shock wave. When it reaches the reflector, the plutonium within is still subcritical. The reflector acts like a piston, slamming inward. The radius of the plutonium decreases. Its density increases. As the shock wave reaches the outer surface of the plutonium, the material is almost exactly at the point of criticality. When the shock wave hits the center, the plutonium has been compressed above its normal density and is supercritical. With plutonium atoms now so dense in the core, the probabilities of collision between free neutrons and the plutonium nuclei are considerably increased. The result is an atomic fireball.

Taylor leaned back and looked up into the sky and seemed for the first time to be aware of the falling rain. He said there was something about the structure of implosion bombs that he had not gone into, and that he could not go into, which contributed greatly to their yield. He said it had proved out in a bomb he had designed, which had had a very descriptive name.

"What was the name?" I asked him.

He shook his head. After a moment, he said, "All I can say is this: They had known all along that the way to get more energy into the middle was to hit the core harder. When you hammer a nail, what do you do? Do you put the hammer on the nail and push?"

"I DON'T think it's likely to be a threat to world peace. It's more a James Bond fear than a real one," said Manson Benedict, emeritus professor of nuclear engineering at M.I.T.

"It's 'Mission: Impossible,'" said Burton Judson, manager of General Electric's Midwest Fuel Recovery Plant, in Morris, Illinois.

"I don't see a radical group coming up to this plant and blazing away with machine guns. I don't. There are nonnuclear alternatives for radical groups. If I was involved with a radical group, I'd go after five hundred pounds of arsenic," said Roger Wiggins, manager of Westinghouse's plutonium-fuel-fabrication plant, in Cheswick, Pennsylvania.

"A self-respecting ambitious terrorist has better things to do than to take nuclear material," said James Schlesinger, when he was chairman of the Atomic Energy Commission.

"Biological and chemical agents are less complex and more available," said Delmar Crowson, when he was director of the A.E.C.'s Division of Nuclear Materials Security.

"Botulism could be used to put this whole city—any city—to an early death," said Leonard Brenner, Crowson's successor.

To a joint congress of the Atomic Industrial Forum and the American Nuclear Society, Thomas Kimball, executive vice-president of the National Wildlife Federation, said, "A large segment of the conservation movement thinks atomic energy is the way to go."

LITTLE BOY, of Hiroshima, was a thirteen-kiloton bomb. It killed nearly a hundred thousand people —a fact later filed under weapons effects. The most densely populated sector of the world is the part of Manhattan Island synecdochically known as Wall Street, where, in a third of a square mile, the workaday population is half a million people. If all the people were to try to go outdoors at the same time, they could not do so, because they are too many for the streets. A crude bomb with a yield of only one kiloton could kill a couple of hundred thousand people there. Weapons effects. Because the tall buildings would create something known as "shadow effect," more than twenty-five kilotons would be the yield necessary to kill almost everybody in the financial district. High dams taper, are thinner at the top. One kiloton would destroy at least the upper half of any dam in the world. Hoover Dam has the biggest head of water in the United States. A bomb dropped behind it into Lake Mead and set to go off at a depth of fifty feet would pretty much empty the lake. Weapons effects. The yield necessary to kill everyone in the Rose Bowl is a fizzle yield, something on the scale of one-fiftieth of a kiloton—so little that it would be not shock or fire but gamma rays that did the killing. A tenth of a kiloton detonated outside an electric-power reactor could breach the containment shell, disable the controls, and eliminate the emergency core-cooling system. There is more long-lived radioactivity in a reactor that has been running for a year than there would be in a bomb of a hundred megatons. A bomb with a yield of a fiftieth of a kiloton exploded just outside the spent-fuel pools at a reactor or a reprocessing plant could send downwind enough strontium-90 alone to kill tens of thousands of people. The placement of an explosion—where it happens—is what matters most, and that depends on purpose. The Hiroshima and Nagasaki bombs were exploded eighteen hundred and fifty feet in the air, because the guess was that from that height the bombs would accomplish the most damage through shock, fire, and radiation effects. A low-yield bomb exploded inside one of the World Trade Center towers could bring it down. The same bomb, if exploded outside, would perform erratically. The Pentagon is a hard target, because it is so spread out. A low-yield bomb exploded in the building's central courtyard would not be particularly effective. To crater the place and leave nothing but a hole in the ground, a full megaton—set off in the concourse, several levels under the courtyard—would be needed. Weapons effects.

A one-fiftieth-kiloton yield coming out of a car on Pennsylvania Avenue would include enough radiation to kill anyone above the basement level in the White

House. A one-kiloton bomb exploded just outside the exclusion area during a State of the Union Message would kill everyone inside the Capitol. "It's hard for me to think of a higher-leverage target, at least in the United States," Ted Taylor said one day. "The bomb would destroy the heads of all branches of the United States government—all Supreme Court justices, the entire Cabinet, all legislators, and, for what it's worth, the Joint Chiefs of Staff. With the exception of anyone who happened to be sick in bed, it would kill the line of succession to the Presidency—all the way to the bottom of the list. A fizzle-yield, low-efficiency, basically lousy fission bomb could do this."

The Massachusetts Turnpike, as it bisects Boston, passes directly underneath, right through the basement of, the Prudential Center, a building complex that includes a fifty-two-story skyscraper. "All you'd have to do is stop, lift the hood, and beat it," Taylor noted as we drove through there one day. We went up to the top of the building to view the city. After a long look and a long pause, he said he could not imagine why anyone who went to the trouble to make a nuclear bomb would want to use it to knock over much of anything in Boston.

Driving down from Peekskill, another time, we found ourselves on Manhattan's West Side Highway just at sunset and the beginning of dusk. There ahead of us several miles, and seeming to rise right out of the road, were the two towers of the World Trade Center, windows blazing with interior light and with red reflected streaks from the sunset over New Jersey. We had been heading for midtown but impulsively kept going, drawn irresistibly toward two of the tallest buildings in the world. We went down the Chambers Street ramp and parked, in a devastation of rubble, beside the Hudson River. Across the water, in New Jersey, the Colgate sign, a huge neon clock as red as the sky, said 6:15. We looked up the west wall of the nearer tower. From so close, so narrow an angle, there was nothing at the top to arrest the eye, and the building seemed to be some sort of probe touching the earth from the darkness of space. "What an artifact that is!" Taylor said, and he walked to the base and paced it off. We went inside, into a wide, uncolumned lobby. The building was standing on its glass-and-steel walls and on its elevator core. Neither of us had been there before. We got into an elevator. He pressed, at random, 40. We rode upward in a silence broken only by the muffled whoosh of air and machinery and by Taylor's describing where the most effective place for a nuclear bomb would be. The car stopped, the door sprang back, and we stepped off into the reception lounge of Toyomenka America, Inc., a Japanese conglomerate of industries. No one was behind the reception desk. The area was furnished with inviting white couches and glass coffee tables. On the walls hung Japanese watercolors. We sat down on one of the couches. "The rule of thumb for a nuclear explosion is that it can vaporize its yield in mass," he said. "This building is about thirteen hundred feet high by two hundred by two hundred. That's about fifty million cubic feet. Its average density is probably two pounds per cubic foot. That's a hundred million pounds, or fifty kilotons—give or take a factor of two. Any explosion inside with a yield of, let's say, a kiloton would vaporize everything for a few tens of feet. Everything would be destroyed out to and including the wall. If the building were solid rock and the bomb were buried in it, the crater radius would be a hundred and fifty feet. The building's radius is a hundred feet, and it is only a core and a shell. It would fall, I guess, in the direction in which the bomb was off-centered. It's a little bit like cutting a big tree."

In dark-blue suits, in twos and threes, Japanese businessmen came out of the warrens of Toyomenka. They collected at the elevator shaft. In voluble streams of Japanese, they seemed to be summarizing their commercial day. More came, and more. None of them seemed to notice or, certainly, to care that we were there. "Thermal radiation tends to flow in directions where it is unimpeded," Taylor was saying. "It actually flows. It goes around corners. It could go the length of the building before being converted into shock. It doesn't get converted into shock before it picks up mass."

We went down a stairway a flight or two and out onto an unfinished floor. Piles of construction materials were here and there, but otherwise the space was empty, from the elevator core to the glass façade. "I can't think in detail about this subject, considering what would happen to people, without getting very upset and not wanting to consider it at all," Taylor said. "And there is a level of simplicity that we have not talked about, because it goes over my threshold to do so. A way to make a bomb. It is so simple that I just don't want to describe it. I will tell you this: Just to make a crude bomb with an unpredictable yield—but with a better than even chance of knocking this building down—all that is needed is about a dozen kilos of plutonium-oxide powder, high explosives (I don't want to say how much), and a few things that anyone could buy in a hardware store. An explosion in this building would not be completely effective unless it were placed in the core. Something exploded out here in the office area would be just like a giant shrapnel bomb. You'd get a real sheet of radiation pouring out the windows. You'd have half a fireball, and it would crater down. What would remain would probably be a stump. It's hard to say which way the building would fall. It would be caving one way, but it would be pushed the other way by the explosion." Walking to a window of the eastern wall, he looked across a space of about six hundred feet, past the other Trade Center tower, to a neighboring building, at 1 Liberty Plaza. "Through free air, a kiloton bomb will send a lethal dose of immediate radiation up to half a mile," he went on. "Or, up to a thousand feet, you'd be killed by projectiles. Anyone in an office facing the Trade Center would die. People in that building over there would get it in every conceivable way. Gamma rays would get them first. Next comes visible light. Next the neutrons. Then the air shock. Then missiles. Unvaporized concrete would go out of here at the speed of a rifle shot. A steel-and-concrete missile flux would go out one mile and would include in all maybe a tenth the weight of the building, about five thousand tons." He pressed up against the glass and looked far down to the plaza between the towers. "If you exploded a bomb down there, you could conceivably wind up with the World Trade Center's two buildings leaning against each other and still standing," he said. "There's no question at all that if someone were to place a half-kiloton bomb on the front steps where we came in, the building would fall into the river."

We went back to the elevator, and when the car stopped for us it was half filled with Japanese, who apparently quit work later than everyone else in world trade. Thirty-eight floors we fell toward the earth in a cloud of Japanese chatter, words coming off the Otis walls like neutrons off a reflector. In the middle of it all, I distinctly heard one man say a single short sentence in English. He said, "So what happened then?"

IT seems inevitable to Ted Taylor that any civilization in the universe must at some point be confronted with a nuclear crisis, and among all the civilizations ahead of us in scientific development he wonders how many are dead and how many have got through the crisis safely. Passionately, he wonders how the successful ones did it. He refuses to believe that we cannot succeed as well. He thinks that the United States, for its part, is "in the foothills picking daisies and has not yet begun to climb the mountains; a deadline is on us; it is almost too late." A safeguards system going far beyond what there is now and involving everything from Doberman pinschers to satellite communication systems has to be designed and put into effect.

"We rely on our people," someone told us in the course of our travels. "We rely on our people, not on mechanical or instrumental safeguards. We look on strangers with a skeptical eye." In the doorway of a plutonium-fuel-fabricating plant Taylor would prefer something more than a gimlet eye. The West Germans, at a plant in Karlsruhe, have doorway monitors that will sound an alarm if so much as a gram of plutonium comes anywhere near them. Doorway monitors are not required in the United States—gamma-ray counters, neutron counters, metal detectors. Under the new A.E.C. regulations, companies may install them or may substitute a personal search by a guard. Much might be learned from prisons. If four high walls with sharpshooters positioned at the corners are deemed necessary for the containment of felons, why are they not necessary for a hundred bombs' worth of plutonium?

Leonard Brenner, director of the A.E.C.'s Division of Nuclear Materials Security, does not think that transportation is necessarily the weakest link in the civilian power fuel cycle. "It depends on who wants the stuff," he said. "If someone does not care that the theft is overt, then transportation is the most vulnerable spot. If the bomber cares, then he has to pilfer the stuff piecemeal, undetected—and not from trucks. A covert man would have to work in a plant and understand its materials-accountability system. This is why that system needs improving."

Delmar Crowson, Brenner's predecessor, said that what is needed is a computerized system in which all material is constantly measured—providing what he called "real-time data." He said the A.E.C. should have computers that talk to the computers of, for example, Consolidated Edison. "Then if someone says there's something missing, we'd be able to say what it is—ideally—in a few minutes." For people in the nuclear-power business who have access to weapons-grade material, no security clearance is required. The A.E.C. recommends that it be required. Brenner has suggested that, at the very least, periodic psychiatric evaluations might be made of people who deal with special nuclear materials. Acts of Congress are needed for such moves, and Congress has not acted.

Perhaps the Red Army—out of sheer international prudence—would be eager to transport our fissile material for us, since our own Army (bad for the image of the atom) has been considered and rejected for the job. Possibly it would be best to ship everything by rail, under armed guard, and with a transmitter on every canister. Ted Taylor believes that all fissile material in transit should be weighed down with steel and concrete to the point where no gang without a crane could move it. Is it better to send ten shipments of two hundred kilograms or one shipment of two thousand kilograms? He emphatically favors the big shipment—convoyed, superguarded. Should fissile material be labelled as such when in transit? If it is not labelled, it is, in effect, camouflaged in the immense flow of non-weapons-grade material that travels the same routes. The industry, however, has reported to the A.E.C. that enough informants exist to make that particular disguise pointless.

A truck full of plutonium had once begun a journey of hundreds of miles when the shipper realized an error had been made and wanted to call the truck back. Having no radio contact with the driver or any other way of communicating with him, the shipper called state and local police. Would someone please stop the plutonium truck and tell the driver to turn around? The truck made its way through the towns and counties of five states, and the police never found it at all. ("The police area needs a lot of attention," said someone at the A.E.C.) Police, as a whole, know nothing of the sensitivity of the fissile materials that pass through their states and communities. They do not have a sense of the magnitude of the threat, the importance of finding and recovering—perhaps dying in the attempt—what they might be called upon to try to find and recover.

Fuel that is fabricated in Wilmington, North Carolina, may go to California to be fissioned in a reactor, then to Morris, Illinois, to be reprocessed. The distances are typical. The nuclear-power fuel cycle is disjoint. It probably should be required by law that chemical reprocessing plants and fuel-fabrication plants that handle weapons-grade material be built side by side. Then the plutonium nitrate that comes out of the one plant could go straight into the other plant, and on into a form—mixed-oxide fuel pellets—far less readily convertible into a bomb. The long trip as nitrate or pure oxide would be avoided. Meanwhile, plans and applications continue to come forth for reprocessing plants and fuel-fabricating plants in places spread out through the land. Near West Valley, New York, Nuclear Fuel Services has considered building a plutonium-recycle fuel-fabrication plant next door to the ASDA storage facility. This would make something of a nuclear park of the West Valley nuclear reservation, and that is what the State of New York intended when, in 1961, it bought some three thousand acres outside West Valley. The state has envisioned a garden of atoms, with reactors and reprocessing and fuel-fabrication and conversion facilities agglomerate. All components of the fuel cycle could be within one compound, including many reactors—uranium ore coming in one end and electricity going out the other. Opposite the safeguards advantages of such a nuclear park, the disadvantages are the distances required for transmission lines and susceptibility to sabotage.

There is irony in the nuclear-safeguards problem. It is possible that by the end of the twentieth century all the vast burgeoning of plutonium from light-water and breeder reactors—a million kilograms per year—will end. It is hoped and, by some, expected that thermonuclear-fusion reactors will by then start producing electricity. Their fuel will be the heavy isotopes of hydrogen, the most abundant energy resource on the earth, and the energy crisis will cease to be a crisis for many millions of years. The fusion reactor is a paper reactor now, of course, and, as Admiral Rickover has said, "A paper reactor works best." But with fusion there is very much less of the kind of risk that worries Ted Taylor with plutonium-239 and uranium-235. It will be extremely difficult, though perhaps not impossible, to make a bomb with anything having to do with a pure-fusion reactor. But meanwhile enough material will have been produced during the fragile interim—in the final, desperate decades of this six-packed and air-conditioned century—to destroy every urban area in the world several times over.

One helpful decision might be to spend more money on fusion research. It is, as Glenn Seaborg has described it, "the most difficult scientific-technological project ever undertaken by mankind." Seaborg said at the same time—speaking before the Atomic Industrial Forum and the American Nuclear Society—that fusion is safer than fission, for there is no possibility of a runaway accident. "Nuclear power is still invisible compared with its imminent role," he said, and, in passing, he urged everyone to remember that "the synthetic environment is as important as the natural one." The government has spent many times as much money on fission research as on fusion research. Less has been spent on fusion than on research having to do with coal. Much smaller amounts have gone for studies of geothermal energy and solar power. The largest source of energy within twenty-five trillion miles is the sun. The sun is a fusion reactor. Perhaps the sun will eventually replace all reactors built on earth.

The continuing existence of the nuclear-power fuel cycle, for all its problems, seems inevitable. People will live with it in the way that others have lived with fear of the sword. The question is not so much whether it is good or bad as whether mankind can live with the bad part—a bomb now and again going off God knows where—in order to have the good. "Every civilization must go through this," Taylor repeats. "Those that don't make it destroy themselves. Those that do make it wind up cavorting all over the universe." As I followed him around the country, people kept asking me, more often than I can count from memory, if I realized the size of the investment that is already implanted in the nuclear industry and how damaging to that investment any major change in the industry's patterns could be. I concluded that other civilizations may well have died rich.